T5-AFV-236

Proceedings of the Twenty-Ninth
Annual Biology Colloquium

The Annual Biology Colloquium

YEAR, THEME, AND LEADER

1939. *Recent Advances in Biological Science.* Charles Atwood Kofoid
1940. *Ecology.* Homer LeRoy Shantz
1941. *Growth and Metabolism.* Cornelis Bernardus van Niel
1942. *The Biologist in a World at War.* William Brodbeck Herns
1943. *Contributions of Biological Science to Victory.* August Leroy Strand
1944. *Genetics and the Integration of Biological Sciences.* George Wells Beadle
1945. (Colloquium canceled)
1946. *Aquatic Biology.* Robert C. Miller
1947. *Biogeography.* Ernst Antevs
1948. *Nutrition.* Robert R. Williams
1949. *Radioisotopes in Biology.* Eugene M. K. Geiling
1950. *Viruses.* W. M. Stanley
1951. *Effects of Atomic Radiations.* Curt Stern
1952. *Conservation.* Stanley A. Cain
1953. *Antibiotics.* Wayne W. Umbreit
1954. *Cellular Biology.* Daniel Mazia
1955. *Biological Systematics.* Ernst Mayr
1956. *Proteins.* Henry Borsook
1957. *Arctic Biology.* Ira Loren Wiggins
1958. *Photobiology.* F. W. Went
1959. *Marine Biology.* Dixy Lee Ray
1960. *Microbial Genetics.* Aaron Novick
1961. *Physiology of Reproduction.* Frederick L. Hisaw
1962. *Insect Physiology.* Dietrich Bodenstein
1963. *Space Biology.* Allan H. Brown
1964. *Microbiology and Soil Fertility.* O. N. Allen
1965. *Host-Parasite Relationships.* Justus F. Mueller
1966. *Animal Orientation and Navigation.* Arthur Hasler
1967. *Biometeorology.* David M. Gates
1968. *Biochemical Coevolution.* Paul R. Ehrlich
1969. *Biological Ultrastructure: The Origin of Cell Organelles.* John H. Luft
1970. *Ecosystems: Structures and Functions.* Eugene P. Odum

Biochemical Coevolution

Biochemical Coevolution

Proceedings of the Twenty-Ninth Annual Biology Colloquium, April 26-27, 1968

Edited by Kenton L. Chambers

Oregon State University Press
Corvallis, Oregon

ISBN:87071-168-7 LCC: 52: 19235

Printed in the United States of America

Contents

Preface

BIOLOGISTS have long recognized that the evolution of any given organism is affected by an exceedingly complex set of interacting factors. To make headway in their efforts to describe and dissect the evolutionary process, they often arrange the most obvious factors in a simplified classification—physical factors, biotic factors, temporal factors, and so forth. Depending upon the degree of resolution desired, relatively more sophisticated classifications can be developed, as for example the mathematical abstracts of population genetics. But we obviously depend on our ability to analyze factors in this simplified way, because their true complexity is overwhelming. In the broadest sense, the biological attributes of any species are a manifestation of its total evolutionary history and, in fact, of the whole history of life. It is not possible to explain the evolution of a particular structural, behavioral, or physiological adaptation in totality; rather, we hope to develop a reasonable and consistent explanation in terms of presently existing, measurable factors plus a minimum of assumptions about the past history of the species. Our overall knowledge of biological evolution grows, therefore, by small increments; it is conceptually a structure upon which many builders are working simultaneously, but which grows in form and stability as construction proceeds.

The reason for emphasizing the analysis of evolutionary factors, in this preface to a book on Biochemical Coevolution, is that I feel the intellectual appeal of evolutionary biology comes largely from its continuing revelation of the unexpected. Every technical advance in biology opens up new vistas for the evolutionist, and for almost every important new datum an evolutionary context can be found. A major "factor" that is only now being investigated in many cases of biotic co-adaptation is the biochemical one. And what this approach reveals is a newly appreciated level in the adaptive responses of organisms, which is based on their ability to survive in a milieu conditioned by the biochemical products of other living things. Some of the resulting adaptations had been guessed at before and are now being confirmed; others were almost totally unrecognized. With the necessary techniques and knowledge of biochemistry, future students should find an abundance of interesting problems for research in evolutionary biology.

The contributions of all the Colloquium speakers except Dr. Robert W. Hull were available for inclusion in this volume. Omitted were the questions and discussions raised by the audience, since this material was for the most part of minor interest compared to the formal presentations. I am very grateful for the conscientious work of the local committee, and especially Dr. William C. Denison who had to assume completely the duties of chairman of local arrangements due to my being on leave-of-absence during the 1967-1968 academic year.

Kenton L. Chambers

Coevolution and the Biology of Communities

PAUL R. EHRLICH
Department of Biological Sciences
Stanford University, Stanford, California

IN RECENT YEARS ecologists have been focusing more and more attention on the properties of communities of organisms. There has been a renaissance of what we used to call "synecology," and several new schools of community ecologists have emerged. One school has focused its interests on the concepts of niche, species diversity, and related topics. Members of this school often deal with questions such as: "Why are there more species of lizards on island X than on island Y?"; "What simple environmental measures can I use to predict the number of bird species in a grassland?"; or "What limits the similarity of sympatric species?" Another approach to communities which has gained prominence recently is a holistic-mathematical approach. Many measurements are made of a complex ecological system. Then the analytic and simulation techniques of systems analysis are used to identify important variables and predict future states of the system.

This colloquium deals with another new way of looking at the properties of communities. This way consists of examining the patterns of interaction not in an entire community but between two groups of organisms, groups which do not exchange genetic information but which do have a close and evident ecological relationship. Peter Raven and the author (1965) called the evolutionary interactions within such systems "coevolution" in order to emphasize the reciprocal nature of the relationship. This reciprocity is abundantly evident in the butterfly-plant systems which we investigated and in herbivore-plant systems in general.

Plants and herbivores

Many of the characteristics of plants, such as spines, pubescence, nutrient-poor sap, and so-called "secondary plant substances" have evolved in large part in response to selection pressures created by

AUTHOR'S NOTE: This paper is dedicated to my friend, Theodosius Dobzhansky, on the occasion of his seventieth birthday, January 25, 1970.

herbivores. The chemicals seem to be especially important, serving as both repellents and pesticides. Herbivores, on the other hand, have responded to the defenses of plants in diverse ways. Many obviously have adopted detoxifying systems to deal with the noxious compounds produced by the plants. For instance the plant *Lotus corniculatus* occurs in populations polymorphic for the presence of cyanogenic glucosides (Ford, 1964). The plants containing the cyanogenic glucosides produce hydrogen cyanide when they are injured. Not surprisingly, these plants are much less bothered by herbivores than their noncyanogenic cohorts. Some herbivores, however, eat both kinds of plant with equal gusto. One of these is the blue butterfly *Polyommatus icarus*. Lane (1962) suggested that the larvae of the butterfly detoxify the cyanide by converting it into thiocyanate with the enzyme rhodanase.

Some insects have been so successful in dealing with plant poisons that they now recognize and are attracted to compounds which repel most herbivores. Indeed one "herbivore," *Homo sapiens,* consumes large quantities of plants because of the many uses he has found for "plant pesticides." He uses them in the role for which they evolved—as herbivore poisons (e.g., pyrethrum) and as herbivore intoxicants (various hallucinogens) and in roles unrelated to their original purpose (pepper, quinine, tobacco). Perhaps the best all-around response to plant defenses is found in aposematic organisms (those that advertise their defensive abilities by conspicuous patterns and coloration, such as the monarch butterfly). These organisms take up the plant chemical defenses and use them for their own protection. The monarch butterfly, for instance, is avoided by most birds because it contains vertebrate heart poisons. These poisons are obtained directly from the monarch's milkweed foodplants. Monarchs and similar organisms gain additional advantage from the avoidance of their foodplant by other herbivores (Reichstein et al., 1968).

Many herbivores have adopted strategies to avoid plant defenses rather than overcome them. Some, for instance, may feed on parts of the plant which have relatively weak mechanical or chemical defenses. An example of this may be flower-feeding or pollen-feeding by many lycaenid butterflies, bees, and various beetles including scymnine coccinellids, dermestids, cantharids, and so forth. Other herbivores time their attacks carefully to avoid plant defenses. Paul Feeny (pers. comm.) reports that the larvae of the winter moth *Operophtera brumata* will not mature satisfactorily on oak leaves two weeks older than those on which the larvae normally feed. Larval development in nature is completed rapidly, early in the season before the tannins are laid down in the young oak leaves.

Plant-herbivore coevolutionary systems usually involve "selectional races." Strong selection pressure is put on plant populations by herbivores, and any improvement in plant defenses is at a selective premium. Herbivores, in response, must find ways of dealing with the plant defenses or they will starve. The tightness of the situation is exemplified by the winter moth case described by Feeny. The timing mechanism of the moths must be extremely precise to guarantee that the larvae hatch just as the young leaves are appearing. If they hatch too early, they starve before the leaves appear; if they hatch too late, they are defeated by the tannins and other plant defenses. If the oaks can evolve ways of depositing tannins even earlier, or produce other defenses, the moths will lose the race, unless the moths can evolve a way of dealing with the oak's defenses.

Perhaps the most unusual coevolutionary system related to herbivory is that composed of swollen-thorn acacias and obligate acacia ants, brilliantly investigated by Janzen (1966). In this system, ants serve as substitutes for the usual defensive mechanisms of acacias. The ant acacias, for instance, lack the bitter-tasting chemicals which are characteristic of other acacias. The ants live in the swollen thorns of the acacias and feed on specially modified leaf tips. If the ants are removed, the acacias are killed by herbivorous insects. As long as the ant colony persists, the ants attack the herbivores and keep them from eating the acacias; the ants also destroy plant competitors of the acacias.

Predator-prey relationships

A coevolutionary system which is the homologue of the plant-herbivore system is the predator-prey system. Like the plant-herbivore system it, in essence, is a selectional race. The prey is selected for predator-avoidance and in the predator for prey-finding. This system is much more familiar to us than the plant-herbivore system; for some strange reason most biologists seem to have the impression that plants just sit around defenseless, waiting to be devoured! All biologists, however, are familiar with the sharp senses and speed of the antelope, the stealth and fangs of the tiger, the spines of the porcupine, and the eyes and talons of the hawk. This is hardly the place to go into the vast literature on this subject, but I do want to point out that the relationship between predator and prey is all too often viewed as static, in spite of the evolutionary work done on *Biston, Cepaea, Natrix,* and mimetic assemblages. There is every reason to believe that most prey species are continually "evolving away" from their predators, and that the predators are either trying to catch up or get ahead. Extinction may very often be the result of a "win" by either side.

In investigating predator-prey systems, most of the emphasis has been placed on the nature and evolution of devices used by prey to avoid being eaten. There is, of course, a vast literature on protective coloration in both vertebrates and invertebrates. An equally impressive literature is accumulating on biochemical defense mechanisms in arthropods as a result of the work of Thomas Eisner and his associates (e.g., Eisner and Meinwald, 1966). Although some research has been done on predator behavior—such as the classic work on orientation in bats by Griffin—much less work has focused on this side of the relationship. Recently, however, there has been an upsurge of interest in the functioning of predators in general, due in particular to the work of Holling (e.g., 1966) in analyzing the components of predation. Once we understand predation more thoroughly, it should be far easier to investigate the reciprocal aspects of predator-prey systems. Perhaps those most amenable to analysis would be systems involving parasitoid wasp predators and their insect prey. Various components of attack and defense have been analyzed in these systems, but to my knowledge none of them have been approached from a coevolutionary standpoint.

Parasite-host systems

Parasite-host systems are similar to predator-prey systems in that one would expect a continuous selectional "race" between host and parasite. The race would be somewhat different, however. It is advantageous for the host, like the prey, to "escape." But, it is not advantageous for the parasite to kill its host, while killing is advantageous for the predator. The counter argument, that it is not advantageous for a predator to eat too many of its prey either, will not hold water. In the vast majority of cases, we must assume that group selection is not operating and that predators which are effective killers leave more offspring than those which are not. There is anecdotal evidence that individual predators may kill far beyond their individual needs for consumption. More importantly, there is no known evidence that any predatory species except man does not feed to repletion, given the opportunity. Conservation of prey resources, if it occurs, is not through the exercise of altruistic restraint by individual predators. With most parasites, however, restraint is not altruistic. Reproductive or feeding behavior which results in the death of the host all too often results in the death of the parasite as well. The problem of the parasite, then, is somewhat more difficult than that of the predator. It must often take care that it does not overtax its resources—for the individuals which do overtax leave fewer offspring than those which do not.

Host-parasite relationships have been studied evolutionarily in some instances, although the evolutionary response has usually been studied one side at a time. Some of the most widely known examples of evolutionary responses involve man: hemoglobin responses to malaria and thalassemia as a host response and the development of drug-resistant bacteria as a parasite response. One host-parasite system is now being studied extensively from both sides by Dr. J. H. Camin and his associates at the University of Kansas. Working with rabbits and rabbit ticks (*Haemaphysalis leporispalustris*), they have been able to demonstrate an immunity to tick attack developing in the rabbits. Ticks that get on a rabbit after others are already feeding either cannot attach or can take much less blood. Therefore, they produce fewer eggs or none at all. Immunity is temporary, only lasting 10 to 20 days if it is not challenged. The ticks show a circadian rhythm in dropping off the rabbits which causes the ticks to concentrate in rabbit warrens and tends to synchronize the life cycles of the individual ticks. They, therefore, get on the rabbit en masse, rather than a few at a time. Many other aspects of the rabbit-tick system are under investigation by the Camin group now, including the fascinating question of why the rabbits have not evolved a long-term immunity. The circadian rhythm of the ticks is, by the way, relatively independent of the rabbit physiology (entrained by photoperiod), which makes an interesting contrast with the European rabbit flea in which the reproduction of the flea and its transfer from the adult rabbit to the young rabbits is controlled by the hormonal changes in the pregnant female rabbit (Rothschild, 1965).

Many other evolutionary responses have been inferred in connection with host-parasite systems, usually as "adaptations" of the parasite to the host and host responses (of which various immune reactions are the outstanding examples). The intimate relationship between parasites and the hosts which we consider to be "vectors" have also received considerable attention, but little evolutionary study. We know that vectors and vectored tend to occur together at the right time and place, but we do not know in most cases what kinds of selective pressures each places on the other. For instance, are *Wuchereria* populations and mosquito populations engaged in a perpetual dance in which a constant disruptive selection pressure occurs in both populations? Microfilariae tend to occur in the peripheral blood at the time that the mosquito vector is feeding. Presumably the earliest and latest mosquitoes have the smallest chance of picking up the parasites. This could lead eventually to a polymorphism of feeding times in the mosquito population, followed by a similar response in the parasite. Or other selection pressures may make either early or late feeding hazardous, so that directional selection would operate on feeding time. Or, other selective factors may override the effects of parasite infection and maintain a rather rigid mosquito feeding time.

Even in the situations where we have some idea of the evolutionary dynamics, as in the case of sickle-cell anemia in man, we have not been able to examine the entire pattern of coevolution. For instance, although we know that hemoglobin S gives considerable protection against *Plasmodium,* we do not know the entire mechanism of protection or the exact kinds of selection pressures to which the parasites are subjected. The glutamic acid-valine residue substitution which changes hemoglobin A to hemoglobin S results in a 50-fold increase in the viscosity of the hemoglobin. The phagotrophic feeding of the *Plasmodium* is inhibited, as the formation of food vacuoles becomes difficult. Changes occur in the surface of infected erythrocytes, making it possible for the liver to recognize them and remove them. The malaria parasite might evade this defense by evolving a new feeding strategy, perhaps by developing an enzyme to reduce hemoglobin viscosity. Whether there is any trend in this direction is unknown. To my knowledge, there has not been an attempt to compare features of strains of, say, *Plasmodium falciparum,* from areas of high and low Hb^S frequency.

Mimicry

Closely paralleling the host-parasite case would be the coevolutionary interactions involved in Batesian mimicry. The mimic, of course, plays the role of the parasite. Its strategy is to take advantage of the model without destroying it. The model gains nothing—and faces the danger of a "credibility gap" developing in its potential predators. For at some point, if the mimics get too common, most predators will associate only happy experiences with what originally was an aposematic pattern in the model. Such a development, of course, ruins the game for both players. One would expect that the model would evolve away from the mimic at a maximum rate, everything else being equal. It is to the mimic's advantage to maintain a maximum of resemblance to the model, until that critical point mentioned above is reached. Then the advantage becomes a disadvantage—the mimic is conspicuously patterned, but predators now associate that pattern with tastefulness. As a result, selection would tend to move the mimic away from the model into a more cryptic pattern. It is not inconceivable that imperfect resemblances, now attributed to mimicry in the process of being perfected, are quite the opposite. They may represent mimics moving away from the model or mimics in an equilibrium situation between perfect mimicry and cryptic coloration. Development of a polymorphism in which one or more forms are nonmimetic may also be the result of such a reversal of the selective situation.

As Ehrlich and Raven (1965) pointed out, there is no sharp line between Batesian and Müllerian mimicry.[1] In all cases it obviously is of advantage to the Batesian mimic to become distasteful if it is physiologically possible to do so. In butterflies, at least, it appears that the usual source of noxious compounds is plant biochemicals, so that foodplant relationships must play a large role in the evolutionary dynamics of any given situation. A butterfly has several different routes to obnoxiousness open to it. If it occurs on a foodplant which does not produce an appropriate compound, it may switch foodplants. If it is feeding on a plant with an appropriate compound or its precursors, the butterfly may evolve the ability to use the compound or synthesize a noxious compound from precursors. Finally the foodplant of the butterfly may evolve an appropriate compound, which then may be picked up by the butterfly. In the latter case the mimetic butterfly would be involved in a complex of "selectional races" involving the model, the foodplant, and predators. As the foodplant becomes more and more obnoxious, the butterfly must find ways of "breaking even" by avoiding poisoning or "winning" by turning the poison to its own advantage. Predators may simultaneously be undergoing selection for ability to discriminate between model and mimic, and for "resistance" to the obnoxious properties of the model. Of course the presence or strength of such selection will depend on many variables. For instance, in some cases butterflies in a single population may make up such a small proportion of the targets of a single predator that selective influence on the predator will be negligible.

In many ways mimetic assemblages make ideal subjects for the study of coevolution—as has been amply demonstrated by the Browers, Phillip Sheppard, and others. We understand a great deal about them, and yet there are many questions unanswered. For instance, detailed studies of putative Müllerian complexes are needed to answer a variety of questions. One would expect that the various members of the complex would have different effects on predators since they presumably are picking up poisons from different sources. As an example, one Müllerian butterfly complex consists of a *Lycorea* species, presumably feeding on Asclepiadaceae or Apocynaceae, several Ithomiines on Solanaceae, two *Heliconius* on Passifloraceae, and a *Perrhybris* with an unknown foodplant. Ideally, of course, each member of the

[1] In Batesian mimicry a palatable species resembles an unpalatable one (the model); in Müllerian mimicry, two different unpalatable species resemble each other.

complex would give strong and similar reinforcement to all local predators, so that multivalent noxiousness might evolve in various members. It would be particularly interesting if biochemical mimicry could be detected in some of these organisms—that is, two quite different chemical compounds obviously selected to give similar effects in the same predator. Rothschild (1961) has suggested that this occurs with defensive odors.

Although it is clear that, in general, Batesian complexes should evolve toward Müllerian complexes, the fate of Müllerian complexes is less obvious. It would probably be unwise to think of them as stable "end points" of evolutionary sequences. If this were the case, one might picture all of the diurnal Lepidoptera (and perhaps many other herbivores and small predators) in an area eventually being recruited into one large complex. It would really save the memories of the birds, but the birds would not have to remember for long because they would starve to death. Obviously, the larger a Müllerian complex gets, and the more similar the defenses of its members become, the more "profit" accrues to a predator which devises a way of consuming the Müllerian mimics. Thus a large selective premium is placed on a strong stomach, and one would expect predators evolving rapidly to deal with the entire complex. If this happens, the advantages of belonging are reduced and one might expect the complex to break up.

Plants and pollinators

Müllerian mimicry is one good example of mutualism. There are many others, many of which doubtless would provide good materials for coevolutionary studies. Perhaps the most widely studied mutualistic coevolutionary system is that of flowering plants and their pollinators. Relationships in this system range from extremely close and clearly reciprocal to casual and possibly unidirectional. Best known of the "tight" relationships are those of the yucca and the yucca moth and of the fig and the fig wasp. In the latter case, both insect and plant are totally dependent on one another—the relationship is obligate in both directions. A wide variety of intimacy has been revealed in the relationships of bees and Onagraceae by the elegant investigations of Linsley, MacSwain, and Raven (1963). A large number of bee species visiting *Oenothera* were found to be oligolectic, collecting pollen for their larval cells exclusively from plants of that genus. Many others, however, were polylectic, collecting pollen from *Oenothera* and from plants of other genera and families. The tightest relationship discovered was that of the bee *Andrena rozeni* and *Camissonia claviformis.* The plant, which has a flower well suited for bee pollination, presents its pollen and nectar in the late afternoon (it is presumably derived from a morning-opening species). *Andrena rozeni* only gathers pollen in the late afternoon, even though residual pollen is available early in the morning. The mouthparts of *A. rozeni* are elongated, permitting it to extract both nectar and pollen simultaneously, and these are very rarely taken from other plants. Mating likewise takes place at the flowers of *Camissonia claviformis,* the males cruising over them before the first appearance of the females.

Of course, many pollination systems have been studied from the point of view of floral morphology, color, and odor in relation to attracting the proper pollinators: long corolla tubes for hawkmoths, red color for hummingbirds, huge widemouthed nocturnal flowers for bats, chemical attractants in orchids, orchids shaped as females to lure male insects, and so forth. It is not clear in most cases, however, what evolutionary responses in pollinators have been elicited by the vast smorgasbord with which they are presented. It would be a mistake to assume that the response has not been considerable, if subtle. It behooves a pollinator to get the job of feeding done with as little energy and risk as possible. Each pollinator is presumably programmed genetically to respond to a "proper" series of stimuli—an exact odor, shade of color, or shape, or a series of odors, colors, or shapes. Each pollinator has adopted a strategy—in essence specialist or generalist. Similarly, each plant has adopted a strategy. As floras and faunas evolve together, the utilities of the various strategies are going to change. The specialist pollinator may find its food source becoming too rare, or the generalist pollinator may find the competition too stiff at many of its sources. Conversely, the specialist plant may find its pollinator going extinct, or the generalist plant may find it is not getting enough accurate transport. The end result of any of these anthropomorphized possibilities is a choice of "evolve or go extinct." When the behavior patterns of pollinators are more thoroughly understood, we shall appreciate more fully the reciprocal nature of most pollination systems.

Coevolutionary complexes and community studies

It seems appropriate now to discontinue the survey of coevolutionary complexes and return briefly to the question of the consequences of their study for community biology in general. Community biology is concerned with the composition of communities and the dynamics of that composition. Community composition is, in part, determined by physical tolerance limits on the distribution of species. Determining the factors limiting distributions and the ways in which organisms "adapt" to the areas they occupy is the preoccupation of a branch of "physiological ecology." Someone once said that the usual conclusion of a study in this field is the determination that the organism can indeed live where it lives. The question of why those limits exist—that is, why the organisms have not transgressed those limits evolutionarily—is rarely investigated. In some cases the answers may lie in the relationship of the organism to its physical environment. For instance, many butterflies may not have penetrated temperate regions simply because they have been unable to develop satisfactory diapause mechanisms. (This, of course, immediately raises the question of why some species have developed satisfactory mechanisms while others have not.) In other cases the answers probably lie in the area of coevolutionary interactions. The presence of a "winning predator" or the absence of a "beatable" foodplant may limit a herbivore. A model may be, in essence, "chased" by a mimic into an area which the mimic cannot penetrate (perhaps because of the distribution of its foodplant). We do know that mimetic species often extend their range beyond that of the model, ordinarily with a rapid loss of mimetic pattern. However, we do not know whether in any cases the mimetic species is restricted to the range of the model because of mimetic relationships.

Taking a coevolutionary approach to problems of community biology lessens the chance of being seduced into "explanations," such as "competition from X limited the distribution of Y." If the limitation of X is due to Y, then the two usually make up a coevolutionary system. In order to understand the limitation, it is necessary to understand the system. This means that questions about selection, such as were asked earlier, must be posed, and field and laboratory experiments must be carried out to find the answers.

The papers which follow in this volume describe research on coevolutionary systems in which biochemical aspects of the systems are receiving special attention.

ACKNOWLEDGMENTS: This work has been supported in part by grants GB-123, GB-2566, and GB-5383 from the National Science Foundation. The author would like to thank Joseph H. Camin (Department of Entomology, University of Kansas) and Peter F. Brussard, Lawrence Gilbert, Jr., Richard W. Holm, Andrew R. Moldenke, Peter H. Raven, and Michael C. Singer (Population Biology Group, Stanford University) for reading and criticizing the manuscript.

Literature Cited

Ehrlich, P. R., and P. H. Raven. 1965. Butterflies and plants; a study in coevolution. Evolution *18:* 586-608.

Eisner, T., and J. Meinwald. 1966. Defensive secretions of Arthropods. Science *153:* 1341-1350.

Ford, E. B. 1964. *Ecological Genetics.* London: Methuen & Co., Ltd., 335 pp.

Holling, C. S. 1966. The functional response of invertebrate predators to prey density. Mem. Entomol. Soc. Canada, No. 48.

Janzen, D. H. 1966. Coevolution of mutualism between ants and acacias in Central America. Evolution *20:* 249-275.

Lane, C. 1962. Notes on the Common Blue (*Polyommatus icarus*) egg-laying and feeding on the cyanogenic strains of Bird's-foot Trefoil (*Lotus corniculatus*). Entomol. Gaz. *13:*112-116.

Linsley, E. G., J. W. MacSwain, and P. H. Raven. 1963. I. *Oenothera* bees of the Colorado desert. II. *Oenothera* bees of the Great Basin. Univ. California Publ. Entomol. *33:* 1-58.

Reichstein, T., J. V. Euw, J. A. Parsons, and Miriam Rothschild. 1968. Heart poisons in the monarch (*Danaus plexippus*). Science *161:* 861-866.

Rothschild, Miriam. 1965. Fleas. Sci. Am., December 1965, pp. 44-53.

The Role of Allelopathy in the Evolution of Vegetation

CORNELIUS H. MULLER
Department of the Biological Sciences
University of California, Santa Barbara

THE EVOLUTION OF VEGETATION is of paramount interest to ecologists, biogeographers, and students of the various applied plant sciences. In recent decades the subject has received some attention from a few ecologists, genecologists, geochronologists, and even ethnologists. Yet, as a subdiscipline of plant ecology it has, most curiously, never developed much beyond analysis of the distribution of certain life forms, groups of related taxa, or genomes. All kinds of evidence, ranging from paleobotanical through archeological to cytogenetic, have been brought to bear upon questions of the movement of plants in time and space. Yet, no organized effort has been made to address this interest toward what these plants did when they arrived or what happened to the species that preceded them. In short, the vegetation that was has become the vegetation that is, but we do not know how because we have not successfully addressed ourselves to the problem of understanding community evolution.

It seems very likely that the subject of community evolution awaits the statement of a basic mechanism of biotic reaction before its theme can be fully enunciated, just as the concept of "survival of the fittest" was required to permit the development of Darwin's thesis. I believe that we now have a component of that mechanism in the form of allelopathy which will be described and related to the process of vegetational development.

The nature of allelopathy

Allelopathy is a term first introduced by Molisch (1937) in dealing with the effects of plant-produced ethylene upon the growth of other plants. It can be defined as the process in which a plant releases into the environment a chemical compound which inhibits the growth of another plant in the same or a neighboring habitat. Biochemical inhibition is a widespread and very important phenomenon among plants. It sometimes produces extremely startling results and, of equal importance, it is sometimes quite subtle and slow but nonetheless effective. It involves a great diversity of chemical compounds in addition to ethylene with which Molisch was dealing.

Two points should be clarified before we proceed further. First, allelopathy does not supersede or even greatly diminish the importance of competition or any other ecological process. It is, in fact, incumbent upon the investigator to determine experimentally the exact significance of all possible factors before ascribing causal value to any one of them. Second, allelopathy is not universal. Several phenomena have been presented in the literature as being the result of biochemical inhibition without any analysis of other possible causes. Some have proved to be unrelated to any phytotoxic qualities of the plants involved. It is well to remember that the presence of a toxin in a plant is no guarantee that allelopathy will result; there must also exist a mechanism by means of which the toxin enters the environment in sufficient concentration to inhibit another plant. It is not vacillation in belief that causes a researcher to deny one allegation, support another, and decide against a third. Even in this sophisticated age of molecular biology, some plants suppress others by means of shade.

Inhibition by Salvia *and* Artemisia

Let us consider now a few examples of biochemical inhibitory phenomena that constitute allelopathic effect. We have described previously (Muller, Muller, and Haines, 1964; Muller and Muller, 1964; Muller, 1966) the means whereby *Salvia leucophylla* and *Artemisia californica* inhibit the seedling herbs of adjoining annual grassland in coastal valleys of Southern California. Thickets of these shrubs gradually invade vast areas of grassland, the inhibited border zones being particularly likely to harbor shrub seedlings. The mechanism of inhibition and consequent invasion is a very complex one, involving not only toxins produced by the shrubs but also several mitigating or intensifying physical factors through which the toxins must operate.

Salvia and *Artemisia* are highly aromatic shrubs which produce large quantities of essential oil comprising numerous terpenes. Although these are known in considerable detail for some species of *Salvia,* we need concern ourselves with only a few of the most copiously produced compounds. In *Salvia leucophylla* we have confirmed the presence of α-pinene, camphene, β-pinene, cineole, and camphor in the fresh foliage and inflorescence. Of these, cineole occurs also in

Artemisia californica together with several other terpenes, most of them unidentified.

We have devised a bioassay technique especially designed to avoid any but aerial contact between the material being tested and the living assay plant. This assay detects only volatile toxins and has revealed the foliage of the aromatic shrub species to be highly toxic to all assay species employed. In early experiments we used *Cucumis sativus* as an assay plant. More recently we have used as assay plants several species of introduced annual herbs that are prominent constituents of the inhibited grasslands under study. *Bromus rigidus* and *Avena fatua* have proved particularly sensitive assay species, and *Bromus rigidus* has become our standard.

Several variants of the basic bioassay method are designed for other purposes. In one, a filter paper seedbed is placed directly upon plant material to permit aqueous contact and to test for water-soluble toxins. In another, an aqueous extract of plant material is used to irrigate the seedbed. Assayed by these techniques, the roots of *Salvia* and *Artemisia* proved not significantly toxic to any of the several herb species employed. The potential toxic effect of these shrubs, then, must involve the volatile terpenes in their foliage.

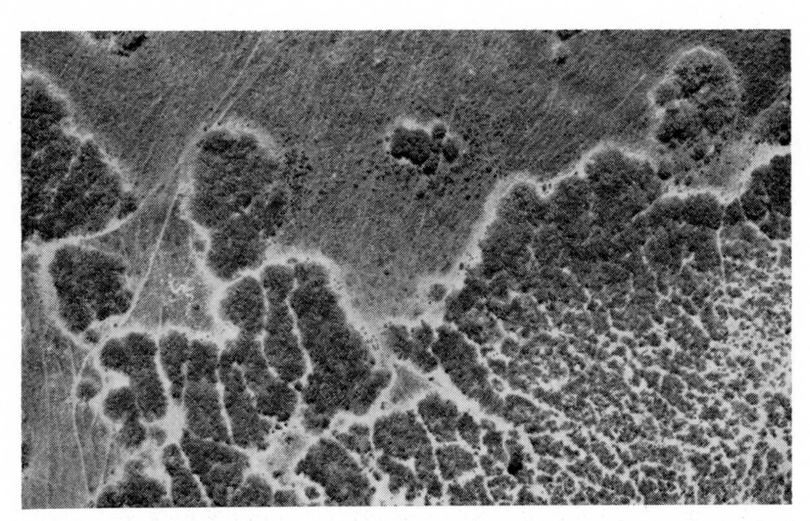

FIGURE 1. Vegetation pattern resulting from the invasion of annual grassland by *Salvia leucophylla* and *Artemisia californica*. Although confined to calcareous clays, this pattern occupies hundreds of square miles in the valleys of the coastal region of Southern California.

If we examine the distribution patterns of the several annual herb species composing the grasslands being invaded by aromatic shrubs, we find strong tendencies for annual plants to be absent under or between the shrubs, even if bare ground is exposed. Extending out from the foliar crowns of the shrubs, there characteristically develops a zone, bare of any herbs, that may be 1 to 2 m in width and frequently exceeds the lateral extension of the root systems of the shrubs. Beyond the bare zone is encountered a gradational zone of herb inhibition, the plants of any one species becoming progressively larger with increasing distance from the shrubs. In this inhibition zone, moreover, some common grassland species are totally absent, occurring only in the uninhibited grassland 6 to 10 m distant from the shrubs.

The inhibition zone commonly contains *Bromus mollis, Festuca megalura, Erodium cicutarium,* and occasionally some unpalatable forbs. All of these are annual plants and they are a mixture of native species and exotics introduced from the Mediterranean region. Ex-

FIGURE 2. Inhibition of annual herbs by *Salvia leucophylla.* The "bare" zone is free of herbs except for a few stunted seedlings. In the next 3 to 6 meters the more tolerant herb species (*Bromus mollis, B. rubens, Festuca megalura,* and *Erodium circutarium*) gradually increase in stature with greater distance from the source of volatile toxins. Beyond the influence of toxins occur *Avena fatua, Bromus rigidus,* and other annual herbs lacking tolerance.

cluded from this zone are *Avena fatua, Bromus rigidus,* and several native annual forbs. The capacity to tolerate close proximity of shrubs seems not to be correlated with the geographic origin of the herb species. It is important to note that adult perennial grasses and forbs appear to be quite tolerant of the shrubs. In some local situations and during some years, the annual species encroach vigorously upon the shrubs.

Testing the hypothesis of allelopathy

Before one is justified in concluding that the phytotoxins contained by the shrubs are responsible for the distributional patterns just described, it is essential that two conditions be meticulously met. First, it is necessary to test systematically the physical and other biotic factors of the environment to determine the degree of involvement of each. In an ecological situation the several environmental factors may be contributory, neutral, or mitigating relative to any specific process. One is justified in concluding that a factor is limiting only after each of the others has been evaluated. Even then, the limiting factor never operates alone, for it cannot reach the process it limits except in the context of the total environment. Second, the contained phytotoxins do not become factors until they are released into the environment and come into effective contact with the plants they are thought to inhibit. It is therefore mandatory that mechanisms of release, transport, and inhibition be discovered, and that these be demonstrated to be effective in each instance for which allelopathy is hypothesized. With both these experimental conditions fully met, one can finally judge the degree, frequency, and geographic extent of a particular allelopathic plant's toxic effect.

In the instance of the *Salvia* phenomenon, we have analyzed the pertinent environmental factors by observation, measurement, or experimentation, depending upon the need. Shading is negatively correlated with the inhibition phenomenon to the extent that light is involved at all. A contact zone between shrub and grass areas usually exhibits inhibition along its total length, but if a portion of this is shaded by an overhanging tree, herbs tend to invade the contact zone and to obliterate the inhibition pattern. Any area that receives full morning sun exhibits the phenomenon more strongly than one getting only afternoon sun. There is no difference in incidence of solar radiation between a bare zone and an adjacent grass zone, and thus light cannot be involved in the growth differences. More likely the action of solar radiation involves temperature and consequent drought stress through increased evaporation. We have repeatedly observed that increased soil moisture reduces the intensity of inhibition, regardless of whether it results from conservation through shading and reduced evaporation or from increase in initial supply (either experimental or as a result of maximal rainfall effectiveness). Yet, no degree of increase in soil moisture has eliminated the inhibition pattern from all or even half its initial area.

Seed germination, in the area of Southern California where *Salvia* was studied, is synchronous for all the herb species involved and occurs quickly following the initial significant rains in early winter. The full process of seedling establishment is accomplished while the soil is uniformly saturated in all zones. Furthermore, the pattern of bare area adjacent to grassland is developed during the process of seedling establishment. The bare zones characteristically produce less than 1 percent as many seedlings as the adjacent grassland. These initial differences in density initiate the pattern without the involvement of any differences in soil moisture. Short periods of slight drought stress occur during the early weeks of growth and these intensify the zonal differences, as will be described shortly.

Trenching across areas of shrub-grassland contact has revealed a high degree of uniformity of soil texture, and analyses for mineral nutrient content were similarly uniform in result. No soil differences correlated with the spatial pattern of vegetation emerged.

The lateral extension of shrub roots was found to reach approximately to the limit of the bare zone, but the soil volume was very poorly occupied and there was no extension of shrub roots into the inhibition zone of stunted herbs. Moisture depletion by shrubs is thus not a cause of herb inhibition. In fact, soil moisture of the upper horizons is demonstrably higher at mid-season in the bare zone than in grassland.

Small animal depredations have been fully investigated and were found capable of significant contribution to expression of the zonal patterns. Nocturnal rodents are heavily concentrated in and near the shrub thickets. A similar concentration of diurnal birds occurs, but these range well into the grass. A system of baiting with the seeds of the pertinent grass species demonstrated that only rodents took *Bromus rigidus* seed and that birds refused them. Rodents, however, failed to range across the broad inhibition zone to the nearest area of *B. rigidus* growth, and thus they could not have eliminated the seed of that species from the inhibition zone. Nonetheless, seed-gathering animals doubtless contribute to the low density of seedlings in the bare zone. Grazing by sparrows is concentrated upon areas of low plant density and doubtless aids in stunting the growth of seedlings in the bare zone. However, we have repeatedly observed localized areas exhibiting no

FIGURE 3. Invasion of inhibited grassland areas by seedlings of *Salvia leucophylla* and *Artemisia californica*. The adult shrubs inhibit their own seedlings less effectively than those of the grasses, but shrub seedlings are extremely rare within old shrub stands. The grass species, some of which are toxic, restrict establishment of shrub seedlings so that localization of their occurrence in areas of grass reduction by chemical inhibition is common.

evidence of grazing at all but nevertheless characterized by stunting significantly different from the rapid growth of the same species only 1 or 2 m distant in the grassland. The curious localization of feeding by flocking sparrows may cause these birds to exert heavy pressure on one bare zone while completely neglecting another only 100 m away. The stunting that occurs free of grazing pressure must be ascribed to proximity to shrubs since it far exceeds the effect of drought stress on windy ridge tops. It occurs, furthermore, during periods of continuing favorable soil moisture such that even densely crowded seedlings in uninhibited grassland show no leaf curl or other signs of water shortage. The role of animals is thus limited, and that of drought has not yet begun when the inhibition pattern is fully established.

Although the shrub species under consideration are evergreen, they become almost fully dormant during the drought of summer and lose most of their leaves. With the advent of winter rains, they resume growth much more slowly than do the herbs and reach full leaf about the time the grasses have completed vegetative growth. Thus the shrubs reach their maximum capacity for terpene production after herb growth is finished. During the early summer drought, we have demonstrated by means of cold traps and gas chromatography that the atmosphere adjacent to *Salvia* shrubs contains cineole and camphor, the two terpene constituents of *Salvia* that proved most toxic in assays of reagent terpenes known to occur in *Salvia*. In experiments involving volatilized terpenes from *Salvia* foliage, it has been demonstrated that dry soil has the capacity to adsorb these gases very rapidly. Similar experiments, furthermore, showed them to be quickly adsorbed by powdered parafin. From these facts, we have devised an hypothesis detailing the mechanism whereby the toxins are effective several meters beyond the reach of shrub branches.

The maximum release of terpenes involves volatilization from mature leaves during the late spring and early summer when rains have ceased and the surface soil is dry. The gaseous terpenes are heavier than air and spread along the soil surface where they are adsorbed upon the colloids of the soil. The soil biota is quiescent at this time so that degradation of terpenes is minimal and oxidation would already have proceeded as far as might be expected. Although high temperature can drive the terpenes off the colloid surfaces, there is little increase in temperature after the period of deposit. With the advent of the rainy season, the terpenes are still present and are in contact with the epidermal cells of seedlings whose radicles grow in contact with soil particles. The ubiquitous cuticular waxes are at least as efficient as soil colloids in accepting terpene molecules, and the hydrophobic terpenes are thus apt to leave the now moist colloid surfaces and pass into solution in the cuticular waxes. The continuity of a lipoidal pathway from the cutin to the rich concentration of phospholipids of the plasma membranes is assured by the wax channels which traverse the epidermal cell wall. This network of lipid concentration extends throughout the seedling body and provides ready transport by diffusion for the highly lipophilic terpenes. Thus the terpenes reach the loci of the growth processes they inhibit.

This inhibitory mechanism of *Salvia* and *Artemisia* shrubs strongly modifies the composition of vegetation in areas invaded by these species. Allelopathy in this instance may be credited with limiting the growth of several species differentially, while apparently not affecting others.

Further examples of allelopathic species

In more recent studies we have concerned ourselves with a wide variety of other inhibitory species, some of which produce water-soluble, nonvolatile toxins. One group of species that exhibits obvious allelopathic patterns comprises the chaparral shrubs of coastal Southern California. There is a conspicuous flush of growth of annual herbs in the first and second growing seasons following a chaparral fire. This is accompanied by the simultaneous establishment of numerous seedlings by some shrub species and the crown-sprouting of others. In the second and third years the native annual species cease germination and disappear, while weedy exotic herbs increase in numbers. The shrub species are meanwhile gradually regenerating typical chaparral thickets. As the shrub cover increases, the density of all annual herb species decreases until these are virtually eliminated at the end of five or six years. Perennial suffrutescent species may last longer, but these also are eventually excluded. In short, seed germination of some native herbs occurs only once, immediately following a burn. Other native and many exotic weedy species continue to germinate for several years. All are eventually inhibited from further germination, and the seeds of most of these lie dormant but viable until the next fire, perhaps 40 years hence.

That the inhibition of germination is allelopathic in nature is strongly suggested (Muller, Hanawalt, and McPherson, 1968) by the absence of herbs from open stands of low shrubs with ample areas of unoccupied soil. Light is maximal in such an area and soil moisture is uniformly at saturation for long periods during the winter rainy season when seed germination normally occurs. Yet no seedlings appear, although adjacent disturbed areas, such as roadsides, produce copious growths of annual herbs each year.

We have determined (McPherson and Muller, 1969) that the allelopathic agent of one of the chaparral shrub species is produced by the shrub foliage. If rain drip is trapped as it drops from leafy branches of *Adenostoma fasciculatum,* it can be demonstrated to be inhibitory to germination and seedling growth of several native as well as exotic species. Furthermore, if the crowns of *Adenostoma* are cut at the soil surface without any soil disturbance, the resulting cleared area shows copious growth of some 30 species of both herb and shrub seedlings in the following growing season.

Other clearly allelopathic shrubs of the chaparral are *Arctostaphylos glauca* and *A. glandulosa.* Fresh leaf litter of these shrubs has been shown to contain several highly toxic phenolic compounds (Hanawalt and Muller, unpub.). Soils occupied by both *Adenostoma* and *Arctostaphylos* have, furthermore, been demonstrated to be toxic to several species of herb seedlings.

Salvia mellifera and *Lepechinia calycina* are both highly aromatic species common in the chaparral. Their terpene content is similar to that of *Salvia leucophylla,* and each has been shown to be highly inhibitory in bioassays. *Salvia mellifera,* furthermore, occasionally occurs in pure stands adjacent to grassland where it produces the same pattern of herb inhibition exhibited by *Salvia leucophylla. Prunus, Heteromeles,* and *Umbellularia* are additional aromatic chaparral associates whose toxicity has been demonstrated *in vitro,* but these have not yet been studied further.

Widespread allelopathic potential thus seems to characterize the dominant shrubs of chaparral, and we are under no illusion that we have found them all.

Among the naturalized exotic species of Southern California, several annual herbs exhibit field evidence of capacity to inhibit the germination and survival of other plants. Notable among these are *Avena fatua, Lolium obtusiflorum,* and *Brassica nigra. Avena* dominates broad areas of grassland from which it excludes numerous species. *Lolium,* after being seeded upon burned chaparral slopes, flourishes for a time at the expense of native herbs and shrubs and then succumbs to its own accumulated toxins. Both of these grasses have been shown to contain toxic phenolic compounds which are leached in quantity from dead plants of the previous season, but there is also strong evidence for effective release of toxins by vigorous young plants (Naqvi and Muller, unpub.).

One of the strongest allelopathic agents among naturalized exotic species is *Brassica nigra.* This annual forb invades grassland most freely in disturbed areas and is only slightly limited in growth by toxic grasses such as *Avena.* When *Brassica* achieves moderate density, it produces a thicket of slender erect stalks which persist for about two years after seed dispersal and death. Although these may reduce light intensity to 50 percent, this is still far above the minimum for lush growth of *Avena, Bromus,* and other annual grasses. An area of *Brassica* recently occupied by grasses and still immediately adjacent to a healthy grass stand is assured a copious supply of grass seed. With the advent of winter rains, the soil is uniformly saturated and seed germination occurs simultaneously in all zones. However, although grass zones are copiously covered by grass and forb seedlings of numerous species, the old *Brassica* thickets produce only *Brassica* seedlings. This inhibition of grass seed germination is often fully developed within 1 or 2 m of normal grassland. It has been determined (Bell and Muller, unpub.) that leachates of the residual *Brassica* stalks are highly toxic and that grass seedlings are much more susceptible than those of *Brassica* itself. It is significant that stalks discolored by saprophytic fungi are many times as toxic as clean stalks. That an obvious instance of allelopathy should ultimately involve a microbial metabolite is no great surprise, but this does emphasize the potential intervention of microorganisms in what appears to be a direct relation between higher plants.

Metabolic wastes as the agents of allelopathy

I have previously supported the hypothesis of de Candolle (1832) that the toxic compounds of plants involved in allelopathy, as well as in the repulsion of pathogens, insects, and browsing animals, are primarily metabolic wastes (Muller, 1966, 1967). The understanding of this phenomenon, like that of many others, has long been rendered difficult by the human compulsion to oversimplify. Animal bodies have prominent muscular devices by means of which they eject metabolic by-products and avoid autointoxication. Because plants lack any such structures, it has been widely held that they do not excrete. The obvious impossibility of sustained metabolism without the production of toxic by-products, and therefore the necessity that plants relieve their active tissues of toxins, seems not to have been taken into consideration. Yet, the literature is full of both direct and indirect evidences of excretion of an extremely wide variety of organic compounds, including carbohydrates, terpenoids, phenolics, alkaloids, amino acids, and even nucleotides (Tukey, 1966; Lundegårdh and Stenlid, 1944). Quite clearly the leaky quality of the plant body permits both volatilization and leaching of these compounds.

Plants that produce highly toxic chemicals necessarily develop some form of either tolerance or protection, else there could be no accumulation of such products without damage. A likely mechanism involves a measure of tolerance combined with some degree of isolation or compartmentalization. For instance, many *Prunus* species contain large quantities of amygdalin. In the same tissues occurs emulsin, a complex of enzymes that reduce amygdalin to HCN and benzaldehyde, both of which are potent phytotoxins. When wounding or senescence breaks down the compartmentalization of amygdalin and emulsin, significant quantities of the toxic materials are released.

If a plant product not only has the power to damage the tissues that produce it but also can resist an invading pathogen, repulse the attacks of insects and browsing animals, or inhibit the growth of competing plants, it then has both deleterious and beneficial potential significance. If, as a consequence of either a small measure of tolerance or an effective isolation within the tissues, it is accumulated without damage to the plant, it may protect against pathogens and browsing animals. If it is excreted from the plant body, its elimination frees the plant of the danger of immediate autointoxication. Its accumulation in the environment, furthermore, works to the detriment of any competitor that is less tolerant to the toxin than is the plant that produces it.

The course of evolution has, of necessity, produced these phenomena one by one. Initially, metabolism produced the toxins whose significance at that point was their power to poison the system in which they appeared. They had to be dealt with either by elimination or by compartmentalization so as to reduce their concentrations to a level tolerable to the system. If this level then proved beyond the tolerance of attacking pathogens or browsers, a secondary significance emerged. If elimination resulted in inhibition of a potential competitor, another form of secondary benefit to the producing plant accrued. In either case, the primary significance was avoidance of autointoxication, for, without this, the system could not have persisted long enough to inhibit anything except itself. Doubtless the selective advantage of secondary benefits, such as reduction of animal depredation, loomed large in the intensification of toxin accumulation once the plant embarked upon a specific evolutionary route of metabolite elimination from its protoplasm.

Autointoxication does occur in plants. The phenomenon of senescence of leaves is accompanied in some instances by obvious accumulations of phenolic compounds which are responsible for the toxic potential of the litter of some woody species. The accumulation of similar products in the vascular tissues of most trees is fatal to these tissues and is circumvented only through regeneration by the annual growth increment of xylem and phloem. Even excreted toxins have been credited with inhibiting the plants that produce them, notably in *Bromus, Helianthus, Hieracium,* and *Salvia* (Benedict, 1941; Curtis and Cottam, 1950; Gouyot, 1957; Muller, 1966).

The stresses that the biochemical sorts of plant reactions place upon individual organisms in the same habitat are comparable in many ways to the stresses of variable temperature, moisture, mineral nutrients, light, and so forth. Another variable in the environmental complex is created to which each organism must be adapted or suffer detriment. Differential tolerances to these biochemical variables have arisen which do not differ in ecological significance from the spectrum of adaptations to shading, drought, and other habitat factors, whether these are biotic or physical in origin.

Correlations with climate and geography

A great many physical factors of the environment doubtless exist which have influenced the pattern of geographic distribution of toxic variables. For instance, any factor that determines the density and vigor of a plant will influence the concentration of its chemical product in the environment. Tolerance limits of species determine where such plants may exist. Let us, however, examine one relation of climate to toxins[1] which has broad geographic implications. It is widely recognized that the semiarid portions of a continent exhibit an inordinate concentration of highly aromatic plants. Whether measured as numbers of species, numbers of individuals, or proportion of the biomass, the vegetations of such climates contain a markedly greater quantity of aromatic plant material than do those of the subhumid and humid climates of the continent. The aromatic quality derives from the large quantity of volatile terpenes these plants produce as metabolic by-products. Conversely, the more humid climates support vegetations rich in species whose metabolic by-products are phenolic in nature.

Terpenes are readily volatilized into a hot, dry atmosphere. They are basically insoluble and immiscible in water. Their hydrophobic quality renders them nonleachable and difficult to eliminate by vaporization through even the thinnest of hygroscopic films that might occur on the surfaces of leaves in a humid atmosphere. Heavy producers of terpenes would therefore be at a distinct disadvantage in the more humid climates.

The phenolic compounds, on the other hand, are water soluble and readily leachable from leaves by rain. In an arid climate the low frequency of leaching renders these compounds difficult to eliminate and operates as an unfavorable factor in the environment of any species requiring leaching. It is significant that the great accumulations of tannic acid exhibited by some arid zone plants are localized in deciduous organs, such as the maturing fruits of some leguminous trees and shrubs.

Thus, from the broad climatic variable of frequency or infrequency of precipitation, there derives a corresponding variable in the proportions of the basic kinds of metabolic by-products produced by plants. The exceptions to this pattern develop compensating adaptations such as the localization of terpenes in the resin canals of conifers in humid climates and the localization of phenolic compounds in deciduous organs of arid-zone legumes. Many other such adaptations exist, most of them totally unexplored. It is impossible to know, of course, how many species that might otherwise invade one or the other end of this climatic spectrum, despite having the wrong sort of metabolic wastes, fail actually to do so for lack of a compensating adaptation.

[1] I am indebted to Dr. R. H. Whittaker for having crystalized my concept of this relation on which I had long been speculating.

A portion of this phenomenon is illustrated by *Eucalyptus globulus* and *E. camaldulensis* in Southern California. Fog drip from *E. globulus* contains phenolic compounds (del Moral and Muller, 1970), while that from *E. camaldulensis* contains none. The leaves of *E. globulus* are readily wettable and therefore adapted to the moist climate of its native Tasmania. Those of *E. camaldulensis* are not leachable until senescence when they fall from the tree loaded with toxic phenolic acids. The riparian distribution of *E. camaldulensis* in dry regions of Australia would make adaptation to leaching a futility.

The relation to competition

My introductory remarks stated that the concept of allelopathy does not diminish the significance of competition as an ecological process. Indeed, every thorough study of an allelopathic phenomenon is a study of possible competitive mechanisms to determine that not these but the toxic mechanism is the limiting factor responsible. One could wish that studies of competition were equally as broad. Plant interference, whether competitive or allelopathic in nature, is doubtless responsible for the exclusion of many species from areas whose qualities would otherwise meet the needs of these species. Since both competition and allelopathy are such powerful determinants of association between species, it might be well to distinguish them definitively. This I have already done more fully elsewhere (Muller, 1969) and only the definitions need be repeated here. We may define interference as *all forms of reaction by one plant that prove deleterious to another.* Competition is a particular form of interference which may be defined as *the process in which the reaction of a plant upon the habitat reduces the level of some necessary factor to the detriment of some other plant sharing the same habitat either simultaneously or sequentially.* The remainder of interference is biochemical in nature and constitutes allelopathy which is *the process in which a plant releases into the environment a chemical compound which inhibits the growth of another plant.* The depleting reaction upon which competition rests is highly nonspecific. Insolation may be reduced to 5 percent of full sunlight by a wide variety of trees which thus produce identical light environments. The chemical qualities of excretions causing allelopathy, on the other hand, show a much higher degree of specificity. While several plants may be comparable in capacity to deposit toxins in the environment, they are usually quite variable in the kind of toxin they liberate. Thus they may produce a spectrum of chemical environments to which associated species will be diversely susceptible or tolerant.

The range of tolerance for a particular toxin exhibited by several plant species may include total susceptibility on the part of some and total tolerance or even stimulation in others. Rice and his students, studying succession in abandoned fields in Oklahoma, found numerous pioneer weeds such as *Sorghum halepense, Helianthus annuus,* and species of *Bromus, Ambrosia,* and *Euphorbia,* which produce phenolic compounds inhibitory to nitrogen-fixing and nitrifying bacteria as well as to their own seedlings. They were not, however, inhibitory to *Aristida oligantha* which forms the next stage of succession (Rice, 1964, 1965a, 1965b, 1965c; Floyd and Rice, 1967; Abdul-Wahab and Rice, 1967). In our own work with *Salvia leucophylla* in California we have noted a higher degree of susceptibility of some herb species than others, a strong tendency of adult plants to be more tolerant than seedlings, and nearly complete tolerance by some adult perennials. From this, one is led to regard terpenes as more active in inhibiting seed germination and the growth of seedlings, particularly of certain species, than the functions of fully established plants.

In the light of these findings, it is not difficult to understand how adjacent areas of similar edaphic and topographic qualities, covered by forest trees casting similar shade, and showing no significant soil moisture or mineral differences can nonetheless exhibit extreme differences in ground cover of herbs. An observant amateur gardener in Maryland remarked to me nearly 30 years ago that his flowering shrubs and herbs grew well under white oak trees, but many of them failed completely under black oaks in the same garden.

Allelopathy and community evolution

I have stated that we have not successfully addressed ourselves to the understanding of community evolution. This has not been for lack of effort nor for lack of genius in some of our elders. It clearly stems from the state of ecological information at that time, which was especially lacking in physiological and biochemical areas.

The first, and perhaps the greatest, step forward consisted of the full enunciation of Gleason's "individualistic concept of the plant community" (Gleason, 1939), which was some 20 years in maturing. Gleason's emphasis upon the survival of selected individuals in a continuously changing environment served as the basis of the continuum concept of Curtis and McIntosh (1951), the development of which is fully reviewed by Whittaker (1967). The mechanistic qualities of Gleason's ideas relieved plant communities of the mystical aura that had long characterized them and still persists in some circles. In this new atmosphere the concept of autosuccession following disturbance in extreme environments was developed (Muller, 1940, 1952; Shreve, 1942). It gave impetus also to studies in genecology, modern or physiological autecology, and allelopathy. Plant ecology, which was born amid folk observations beginning in antiquity, came of age as a science capable of experimental investigation with the growing acceptance of a mechanistic way of thought. Only under such circumstances could a theory of community evolution develop at all.

In an early treatment of this subject, Mason leaned heavily upon the individualistic concept of Gleason in which he made more explicit the involvement of plant reaction upon the environment (Mason, 1947). It fell short of completion, however, largely because the time was not yet right to give full characterization to interference upon which much of plant association as a process depends. My own first effort emphasized the primacy of process in biotic association rather than structure and identity of communities (Muller, 1958). Slight as this presentation of background may be, I intend to give full credit to several of our colleagues for whatever virtue may appear in the statement of the principles of a theory of vegetational evolution now to be attempted.

Under conditions of a variable and ever-fluctuating environment occupied by a diverse flora, still evolving under the selective pressure of habitat, the process of assembling of species is a sieve action by the environment, allocating species to segments of time and space forming local vegetations. The dispersal capacity of plants, although limited in some species and blocked by barriers of widely varying efficiency, places the disseminules of most species in a diversity of habitats. Only a habitat all of whose factors fall within the range of tolerance of an individual will permit its establishment and survival.

Only those species sufficiently close and not barred by a barrier to dispersal can reach a particular site. Gross climatic and edaphic factors determine the general distribution of species and thus their proximity to a site and the availability of their disseminules.

The specific levels of factors encountered by a plant in its habitat are local products of topographic, edaphic, and microclimatic peculiarities of that site as further modified by the reactions of other plants in close proximity. Reaction on physical factors is the basis of competition which may variously suppress or exclude a plant by exceeding its limits of tolerance to a physical factor. Reaction on the organic chemical environment is the basis of allelopathy which may inhibit or exclude a plant by the addition of a toxin to which it is not tolerant or may involve the addition of a stimulant upon which another plant is dependent.

Adaptation of plants to different segments of the habitat avoids competition, while adaptation to modified physical factors accommodates to competition. Possession of tolerance to a specific toxin circumvents allelopathy. Any of these adaptations permits association of a plant with another that might otherwise exclude it either by competition or by allelopathy. Thus, against the background of the original physical environment, plant reaction serves as a fine differential screen, permitting the adapted and excluding the unadapted by competition or by allelopathy.

Reaction upon physical factors is general in nature and common to many species of comparable life form. It is thus uniformly operative over wide areas. Phytotoxins are peculiar to certain species and are operative only within the sphere of influence of those species. A mixture of mutually tolerant dominants may have highly individual allelopathic products and select widely different associates, thus producing diversity of local aggregations of differentially tolerant species.

Environmental flux and continuing biological diversification make all plant assemblages temporary, constituting the vegetation of the moment of a specific habitat. Vegetation is therefore the product of process and *is in process,* capable of further evolution with physical change, speciation, invasion, disturbance, or allelopathy.

Thus, the evolution of vegetation commenced with migration into diverse segments of space by variously adapted species. This process continued as species accommodated to niche differentiation, resulting in reduction of competition. With the addition of biochemical interaction, species were segregated according to the toxic potential of some and the tolerance or susceptibility of others. Continuing genetic change and fluctuation in physical and biochemical environments insures that the resulting vegetation remains a labile assemblage of living systems with a long history of rearrangement, spatially and functionally. Vegetation thus exhibits ultra polyphyletic origin and is capable of continuing its evolutionary development into the future in a reticulate manner.

The omission here of discourse on animal participation does not reflect lack of consideration or even any very great divergence of the principles of animal association with plants or with other animals. Rather, it is likely that each of these processes will be found to function in both groups and that most of them operate also between groups.

ACKNOWLEDGMENTS: This work was supported by the National Science Foundation, contracts GB-149, GB-4058, and GB-6814, and by the Faculty Research Committee, University of California, Santa Barbara. Grateful acknowledgment is made of the collaboration of students and colleagues whose work is cited throughout this paper, and especially to Robert O. Tinnin, reference to whose work on *Avena fatua* was accidentally omitted.

Literature Cited

Abdul-Wahab, A. S., and E. L. Rice. 1967. Plant inhibition by Johnson grass and its possible significance in old-field succession. Bull. Torrey Botan. Club *94:* 485-497.

Bell, D. T., and C. H. Muller. Allelopathic studies of *Brassica nigra.* Unpub. ms.

Benedict, H. M. 1941. The inhibitory effect of dead roots on the growth of bromegrass. J. Am. Soc. Agron. *33:* 1108-1109.

Candolle, A. P. de. 1832. Physiologie végétale. Paris.

Curtis, J. T., and G. Cottam. 1950. Antibiotic and autotoxic effects in prairie sunflower. Bull. Torrey Botan. Club *77:* 187-191.

Curtis, J. T., and R. P. McIntosh. 1951. An upland forest continuum in the prairie-forest border region of Wisconsin. Ecology *32:* 476-496.

del Moral, R., and C. H. Muller. 1970. Fog drip: a mechanism of toxin transport from *Eucalyptus globulus.* Bull. Torrey Botan. Club *96:* 467-475.

Floyd, G. L., and E. L. Rice. 1967. Inhibition of higher plants by three bacterial growth inhibitors. Bull. Torrey Botan. Club *94:* 125-129.

Gleason, H. A. 1939. The individualistic concept of the plant association. Am. Midland Naturalist *21:* 92-108.

Guyot, A. L. 1957. Les microassociations végétales au sein du Brometum erecti. Vegetatio *7:* 321-354.

Hanawalt, R. B., and C. H. Muller. Allelopathic studies of *Arctostaphylos glandulosa* and *A. glauca.* Unpub. ms.

Lundegårdh, H., and G. Stenlid. 1944. On the exudation of nucleotides and flavanone from living roots. Ark. Botan. (Stockholm) *A31:* 1-27.

Mason, H. L. 1947. Evolution of certain floristic associations in western North America. Ecol. Monogr. *17:* 201-210.

McPherson, J. K., and C. H. Muller. 1969. Allelopathic effects of *Adenostoma fasciculatum,* "chamise," in the California chaparral. Ecol. Monogr. *39:* 177-198.

Molisch, H. 1937. Der Einfluss einer Pflanze auf die andere—Allelopathie. 106 pp. Jena.

Muller, C. H. 1940. Plant succession in the *Larrea-Flourensia* climax. Ecology *21:* 206-212.

Muller, C. H. 1952. Plant succession in arctic heath and tundra. Bull. Torrey Botan. Club *79:* 296-309.

Muller, C. H. 1958. Science and philosophy of the community concept. Am. Scientist *46:* 294-308.

Muller, C. H. 1966. The role of chemical inhibition (allelopathy) in vegetational composition. Bull. Torrey Botan. Club *93:* 332-351.

Muller, C. H. 1967. Die Bedeutung der Allelopathie für die Zusammensetzung der Vegetation. Z. Pflanzenkrankh. (Pflanzenpath.) und Pflanzenschutz *74:* 333-346.

Muller, C. H. 1969. Allelopathy as a factor in the ecological process. Vegetation *18:* 348-357.

Muller, C. H., R. B. Hanawalt, and J. K. McPherson. 1968. Allelopathic control of herb growth in the fire cycle of California chaparral. Bull. Torrey Botan. Club *95:* 225-231.

Muller, C. H., W. H. Muller, and B. L. Haines. 1964. Volatile growth inhibitors produced by aromatic shrubs. Science *143:* 471-473.

Muller, W. H., and C. H. Muller. 1964. Volatile growth inhibitors produced by *Salvia* species. Bull. Torrey Botan. Club *91:* 327-330.

Naqvi, H. H., and C. H. Muller. Allelopathic studies of *Lolium obtusiflorum.* Unpub. ms.

Rice, E. L. 1964. Inhibition of nitrogen-fixing and nitrifying bacteria by seed plants. I. Ecology *45:* 824-837.

Rice, E. L. 1965a. Inhibition of nitrogen-fixing and nitrifying bacteria by seed plants. II. Characterization and identification of inhibitors. Physiol. Plant. *18:* 255-268.

Rice, E. L. 1965b. Inhibition of nitrogen-fixing and nitrifying bacteria by seed plants. III. Comparison of three species of *Euphorbia.* Proc. Okla. Acad. Sci. *45:* 43-44.

Rice, E. L. 1965c. Inhibition of nitrogen-fixing and nitrifying bacteria by seed plants. IV. The inhibitors produced by *Ambrosia elatior* L. and *Ambrosia psilostachys* DC. Southwestern Naturalist *10:* 248-255.

Shreve, F. 1942. The desert vegetation of North America. Botan. Rev. *8:* 195-246.

Tukey, H. B., Jr. 1966. Leaching of metabolites from above-ground plant parts and its implications. Bull. Torrey Botan. Club *93:* 385-401.

Whittaker, R. H. 1967. Gradient analysis of vegetation. Biol. Rev. *42:* 207-264.

Invertebrate Symbioses With *Chlorella*

STEPHEN J. KARAKASHIAN
Department of Biology
State University of New York
College at Old Westbury

SYMBIOSES WITH GREEN ALGAE are widely distributed throughout the freshwater representatives of several phyla, including Protozoa, Porifera, Coelenterata, Platyhelminthes, and Mollusca (Droop, 1963; McLaughlin and Zahl, 1966). Among the protozoa and hydra, the algae are definitely known to be intracellular. A typical gastrodermal cell of *Chlorohydra* contains about 10 to 20 algae (Park et al., 1967), while *Paramecium bursaria* may harbor slightly more than 1,000 algae (Karakashian, 1963). Since many of these intracellular symbioses are hereditary, even if not obligate, the potential for evolution of an intimate relationship is obvious.

Among these associations, the algae are remarkably similar to one another and to representatives of the free-living genus *Chlorella*[1]; yet the host species are phylogenetically very diverse. The distribution raises intriguing questions about the evolution of these symbioses. A parallel situation may exist among marine organisms, where a variety of hosts, ranging from protozoa to molluscs, harbor symbiotic zooxanthellae which also seem to be related to one another and to certain free-living dinoflagellates (McLaughlin and Zahl, 1959, 1966; Freudenthal, 1962).

This paper will focus on *Chlorella* symbioses, drawing examples from other systems when they help to clarify the information. The study of alga-invertebrate symbioses has historically suffered from a wealth of opinion and a paucity of critical data. Although this is still

[1] The classification of *Chlorella* species and related forms is difficult and has been the subject of numerous recent publications; a particularly comprehensive study is that of Shihira and Krauss (1963). In stressing the relatedness of symbiotic algae and *Chlorella,* I make no judgment with respect to their precise classification.

true, there has been renewed interest in the subject, and renewed activity, possibly because of a revival of the notion that intracellular organelles may have had a symbiotic origin (Ris and Plaut, 1962; Taylor, 1968).

Information of several types is relevant to a discussion of the origin of these symbioses.

(*1*) The ultrastructure should reflect the degree of morphological divergence between symbiotic and free-living *Chlorella.*

(*2*) Attempts to culture the partner organisms independently should reveal mutual physiological or nutritional dependence.

(*3*) Studies of the cross-infectivity of symbiotic algae into other related or unrelated hosts should clarify the importance of specific adaptations between the partners. Similar studies of the infectivity of free-living algae would be useful.

(*4*) Examination of the functional relationships between the partner organisms in both normal and artificial associations might uncover the mechanisms by which existing physiological adaptations proceed.

Structural features of the symbiosis

The ultrastructure of the *Paramecium bursaria* association has been studied by Vivier et al. (1967) and by Karakashian et al. (1968), and that of the hydra symbiosis by Oschman (1967) and Park et al. (1967). A number of workers have examined the morphology of free-living *Chlorella,* including Northcote et al. (1958), Murikami et al. (1963), Soeder (1964, 1965), Staehlin (1966), and Bryan et al. (1967).

Symbionts from *Paramecium bursaria* and *Hydra viridis* are illustrated in Figures 1 and 2, respectively, and should be compared with the free-living *Chlorella* shown in Figure 3. In general, few ultrastructural features of either the host or algae can be ascribed specifically to the symbiosis. Vivier et al. (1967) have noted, and we have confirmed, however, that the chloroplasts in the intracellular algae of *P. bursaria* are frequently more massive and more compact than those in free-living strains. Preliminary observations of ours indicate that the morphology of a given strain of symbionts may change when cultured outside the host (Figure 4). Accordingly, one must be cautious in ascribing to the genotype responsibility for morphological differences between symbionts and free-living algae unless they have been cultured under similar conditions.

It is of special interest that the cell wall of the symbiotic *Chlorella* shows no evidence of reduction in thickness and structure. In this respect the chlorellae differ from the green flagellate *Platymonas con-*

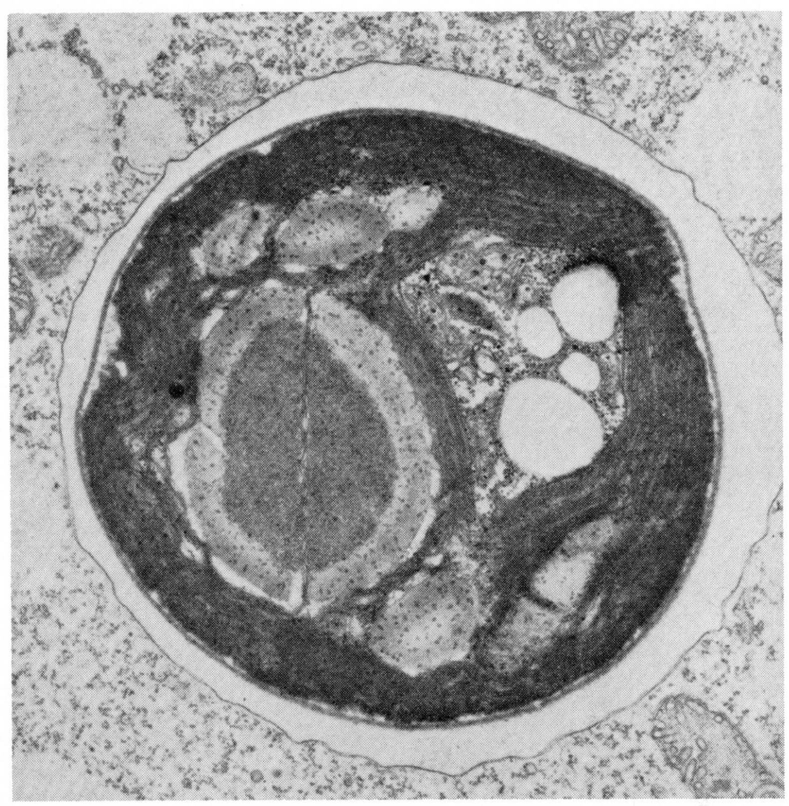

FIGURE 1. Symbiotic *Chlorella* from *Paramecium bursaria* fixed *in situ.* Note the massive cup-shaped chloroplast which is here cut in cross-section, the prominent cell wall, and the host vacuole in which the alga is situated. X 23,500. Osmium fixation.

volutae (Parke and Manton, 1967), which loses its cell wall and becomes almost amoeboid in shape when it infects the flatworm, *Convoluta roscoffensis* (Oschman, 1966). The chlorellae also differ from the blue-green algae which inhabit *Cyanophora paradoxa* and *Glaucocystis nostochinearum,* both of which lack cell walls (Schnepf et al., 1966; Hall and Claus, 1963, 1967). The functional significance of a wall to an intracellular hereditary symbiont is not known, but we have suggested previously (Karakashian et al., 1968) that a massive wall may not be adaptive in the intracellular environment and that selective pressures may tend to reduce it. Accordingly, the existence of a prom-

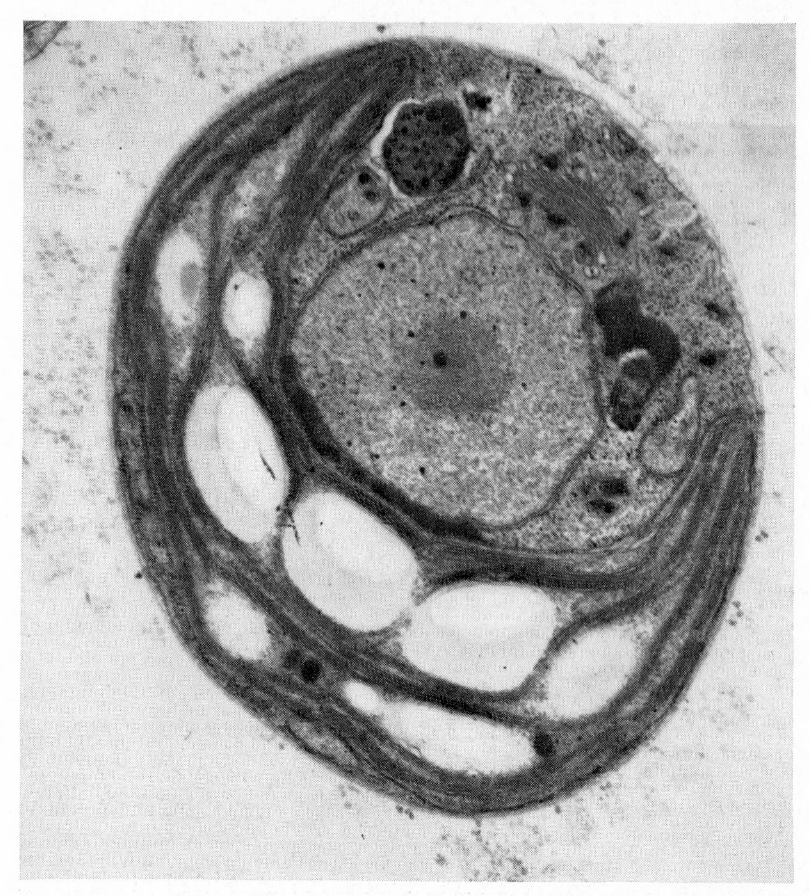

FIGURE 2. Symbiotic *Chlorella* from *Hydra viridis* fixed *in situ*. Note the cup-shaped chloroplast, the large nucleus, dictyosome, and cell wall. X 23,000. Glutaraldehyde-osmium fixation. Image, courtesy of Dr. James Oschman.

inent wall in *Chlorella* could be taken as some evidence that the symbiosis is of recent origin.

A feature common to both *Paramecium bursaria* and hydra is the position of each alga in an individual vacuole. In this respect the natural associations differ from certain artificial ones, since in the latter, numerous algae may be grouped within a single, large vacuole (Park et al., 1967; Karakashian, unpub.). With the normal symbionts, such groups occur only temporarily following division; the vacuole membrane soon invaginates, separating the daughter cells into separate vacuoles once again (Karakashian et al., 1968).

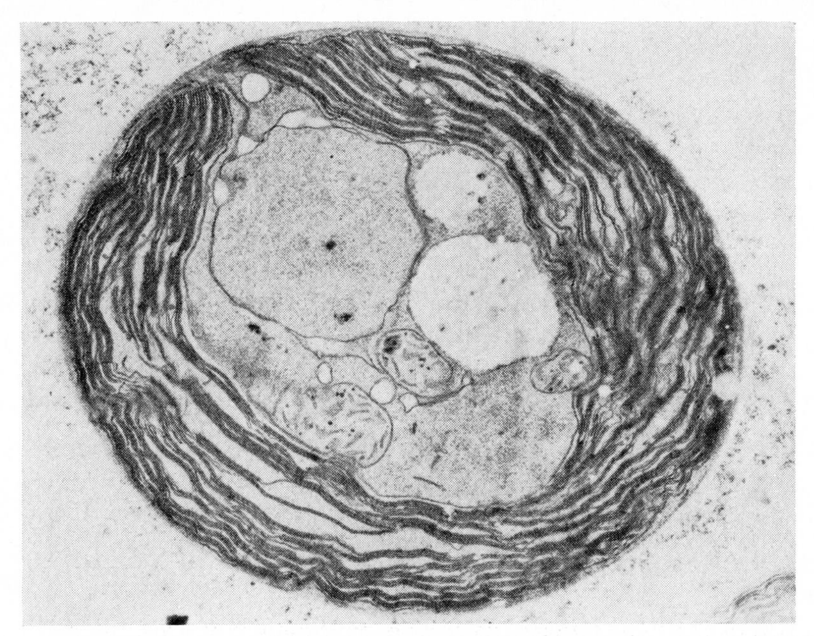

FIGURE 3. Free-living *Chlorella vulgaris,* Granick strain. Note the cup-shaped chloroplast and large apical nucleus. The cytoplasmic matrix and chloroplast stroma appear somewhat less dense than those of the symbiotic algae illustrated because the alga was fixed in $KMnO_4$. X 19,900. Image, courtesy of Dr. Gary Bryan.

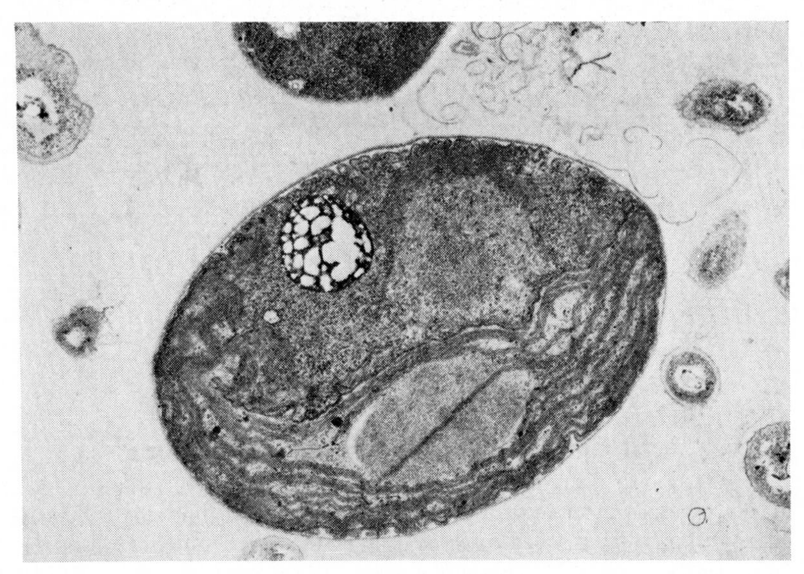

FIGURE 4. Symbiotic *Chlorella* growing outside the *Paramecium* host in the bacterized culture fluid. This alga is from the same clone as the one illustrated in Figure 1. Differences in the morphology of algae growing intra- and extracellularly are evident. X 22,800. Osmium fixation.

Independent culture of the partner organisms

All host stocks, but only some of the algal strains, are capable of independent growth under the conditions tested. Of course, strains which have not been cultured are not necessarily obligately intracellular. However, since most workers have used media which were richly supplemented with organic compounds, it is clear that a relatively high proportion of strains are fastidious (Table 1).

Loefer (1936) cultured algae from *P. bursaria* on a completely inorganic medium, but in our experience most culturable strains will not grow unless organic supplements are provided. The significance of this observation is unfortunately somewhat obscure since the incidence of organic requirements among free-living *Chlorella* strains is not known. Most of the common methods by which free-living strains are isolated employ only inorganic media in order to supress bacterial contamination; accordingly, they select for those strains which lack organic requirements. Nevertheless, considering that a high proportion of symbiont strains have not been cultured and that a sizable proportion of the culturable strains show organic requirements, there is evidence of a developing nutritional dependence of the algae upon the host.

By contrast, the paramecia are always capable of independent growth. The difference, however, may be more apparent than real. The algae, of necessity, must be cultured on a completely soluble medium whose composition must be specified by the experimenter; a single omission may be crucial. The hosts, on the other hand, are invariably fed living organisms, a defined medium not being available for any of them. Under these circumstances, it is no wonder that nutritional dependencies have not been identified. Indeed, in one instance, Loefer (1936) found that algae-free *P. bursaria* would not grow on an axenic medium which was suitable for the growth of infected paramecia.

Infections with free-living algae and algae from other species of hosts

Reports concerning cross-infections with algae from species other than the normal host are fragmentary, and comparisons are therefore somewhat difficult. The reasons why this should be so are not difficult to comprehend.

In order to make a systematic study of interspecific infections, it must be possible to culture the host organisms and to free them of their symbionts. A source of genetically uniform algae should also be available; ordinarily, clones of algae must be cultured. At the least, the in-

Table 1. INDEPENDENT CULTURE OF SYMBIOTIC ORGANISMS

Host	Cultured host?	Cultured algae?	Source
Paramecium bursaria, syngens 1, 6, A, B	Yes (18)	Yes (4), no (5)	Bomford (1965)
P. bursaria, syngen 1	Yes (9)	Yes (7), no (3)	Karakashian (1963 & unpub.)
P. bursaria	Yes	No	Hämmerling (1946)
P. bursaria	No (axenic medium)	Yes	Loefer (1936)
P. bursaria	Yes	......	Oehler (1922)
P. bursaria	Yes	No	Pringsheim (1928)
Stentor polymorphus	Yes	No	Hämmerling (1946)
S. polymorphus	Yes	......	Schulze (1951)
Frontonia leucas	Yes	......	Hood (1927)
Euplotes daidaleos	......	Yes	*Cochrane (pers. comm.)
Spongilla sp.	......	Yes	Lewin (cited in Starr, 1964)
Chlorohydra viridissima	Yes	No	Muscatine (1961 & pers. comm.)

Numbers in parentheses refer to the estimated number of strains of each type tested.
*In Karakashian's laboratory. We thank Dr. W. F. Diller for providing us with the stock of *Euplotes*.

fected hosts must be grown and the algae isolated immediately before use. The disadvantage of this method is that the host may contain a heterogeneous mixture of symbionts whose proportion may vary in unknown ways. Following infection, the hosts must be freed of all extracellular algae; otherwise a true persistence cannot be established. The difficulties attending simultaneous culture of a number of species of hosts and strains of algae have restricted the study of infections, and the possible pitfalls, elementary as they seem, have not always been appreciated by those who have undertaken such experiments. Moreover, different workers have seldom used the same strains, and we cannot assume that all symbiont strains of a given host species will behave identically.

The data which are available show that the pattern of infectivity cannot be predicted from the relatedness of the hosts. The results differ, depending apparently on the specific combinations tested. For example, considering closely related hosts, reciprocal infections among stocks of *Paramecium bursaria,* syngen 1[2], result in permanent infections following exposures of the host to algae for only a few minutes (Siegel, 1960; Karakashian, 1963 and unpub.). Bomford (1965) and Karakashian (unpub.) have shown that algae and paramecia from four different syngens of *P. bursaria* are also mutually compatible. On the other hand, algae from *Stentor igneus* infect *Stentor polymorphus* only with difficulty, despite the fact that both species normally contain algae (Schulze, 1951), and the same is true of *Chlorella* from *Chlorohydra viridissima* when tested with uninfected *Chlorohydra hadleyi* (Park et al., 1967). These data are summarized in Table 2.

Among unrelated host species the situation is similarly variable. We have discovered that algae from three different stocks of *Chlorohydra* readily infect *P. bursaria.* The associations thus obtained are similar to homologous infections with the native *P. bursaria* algae as judged by the stability of the infections, the average number of algae per paramecium, and the ability to stimulate host growth in light (Karakashian, 1963; Karakashian and Karakashian, 1965). By contrast, Hämmerling (1946) and Schulze (1951) found that reciprocal infections between *P. bursaria* and *S. polymorphus* required very long exposures, and the associations were usually unstable when formed. The homologous algae always infected quickly and formed stable associations (Table 2).

A number of workers have attempted to infect free-living strains of *Chlorella* into *Paramecium bursaria* with variable results (Oehler, 1922; Pringsheim, 1928; Siegel and Karakashian, 1959). We under-

---

[2] Syngens are sibling species of the *Paramecium bursaria* complex.

Table 2. INTERSPECIFIC ALGAL INFECTIONS

RELATED HOSTS

Hosts: *Paramecium bursaria*, syngens

Algae:	1	6	A	B
1	+	+	+	+
6	+	+	+	+
A	+	+	+	+
B	+	+	+	+

Bomford, 1965; Karakashian, unpub.

Host: *Stentor polymorphus*

Algae:	*Stentor polymorphus*
S. polymorphus	+
S. igneus	+/–

Schulze, 1951

UNRELATED HOSTS

Hosts:

Algae:	*P. bursaria*	*S. polymorphus*
P. bursaria	+	+/–
S. polymorphus	+/–	+

Schulze, 1951; Hämmerling, 1946

Host:

Algae:	*P. bursaria*
P. bursaria	+
Chlorohydra viridissima	+

Karakashian, 1963, 1965, & unpub.

"+/–" indicates an unstable association or one which formed only following prolonged exposure.

took a thorough investigation of the infectivity of free-living algae, employing 25 different strains of *Chlorella* and related algae and 2 different stocks of *P. bursaria,* syngen 1 (Karakashian and Karakashian, 1965). Representative data are presented in Table 3. A few strains infected neither stock; some infected one stock more heavily than the other; and some infected both stocks. Of the 25 free-living strains tested, 18 showed some infectivity as judged by the persistence for two days of at least a single alga in a nondividing paramecium. Using a considerably more stringent criterion of infectivity, nearly one quarter of the algal strains (6 out of 25) infected at least half the paramecia of one stock or the other, and there was a mean of 50 or more algae per infected paramecium.

It is obvious, then, that a substantial proportion of free-living *Chlorella* strains are infective. In our experiments, a number of non-*Chlorella* algae were slightly infective also, the most successful of which were *Selenastrum minutum* and *Scenedesmus obliquus.* Bomford (1965) reported that a *Scenedesmus* infected *P. bursaria,* and Oehler (1922) described a weak infection with a *Scenedesmus* as well as a highly pathological infection with *Stichococcus bacillaris.* Pringsheim (1928) discovered a similar pathology induced by a strain of *Hormidium nitens.*

We have examined the growth rate of several strains of free-living algae inside the host cells (Karakashian and Karakashian, 1965). A typical case is presented in Table 4. The algae grow much more slowly than the maximum rate of the host, so that, even in light, the average number of algae per paramecium falls rapidly. This finding is surprising since in culture the free-living algae grow more rapidly than do the symbionts. By diluting the medium, the division rate of the paramecia can be slowed until it matches the slower rate of the algae; in this case the partners keep pace with one another. In darkness there appears to be no growth of the intracellular algae, despite the fact that this strain is capable in culture of growth in darkness.

In some preliminary experiments we have asked the question, "Can algae from *P. bursaria* infect related species of paramecia which do not normally harbor algal symbionts?"[3] Although more extensive tests should be made, no instances of infectivity were uncovered.

In this connection, some experiments of Tartar (1953) are of interest (Table 5). He produced interspecific chimeras by grafting together nucleated parts of *Stentor polymorphus* (with algae) and nucleated, enucleated, and partially enucleated parts of *S. coeruleus* (a

[3] Some of these experiments were performed by Mr. Jan Besser in my laboratory.

Table 3. PERSISTENCE OF FREE-LIVING ALGAE IN *P. bursaria*

Algae		Paramecia			
		Stock 42W		Stock 478A-W	
		Proportion infected	Algae/paramecium mean* (maximum)	Proportion infected	Algae/paramecium mean* (maximum)
Chlorella vulgaris	397	0/20	....	0/20	....
	398	12/17	73 (400)	18/20	41 (500)
	263	2/20	1 (1)	19/20	85 (507)
C. pyrenoidosa	395	16/20	204 (633)	4/20	2.5 (5)
Selenastrum minutum	326	18/20	7.7 (17)	9/20	4.2 (10)
Scenedesmus obliquus	72	9/20	27 (209)	10/20	17 (44)
Chlorella sp., (symbiotic)	NC64A	20/20	Range = 510-1,050	20/20	Range = 320-1,060

From Karakashian and Karakashian, 1965.
*Includes only paramecia which were infected.

Table 4. GROWTH OF FREE-LIVING ALGAE IN DIVIDING *P. bursaria*

Algae		Host growth	Estimated initial no. of algae*	No. of algae after 4 fissions of host	Final no. / Initial no.	
					Paramecia	Algae
Chlorella vulgaris	GrB1	Fast, LD	226	504	16	2.2
	GrB1	Fast, DD	324	204	16	0.63
	GrB1	Slow, LD	250	3,290	12†	13
Chlorella sp., (symbiotic)	NC64A	Fast, LD	472	7,058	16	15

From Karakashian and Karakashian, 1965.
*Estimated by counting the number of algae in equivalent host cells.
†Four paramecia died before the experiment could be completed.
"LD" refers to 12 hours of light:12 hours of darkness each 24 hours; "DD" refers to continuous darkness. Growth rates were slowed by diluting the medium.

Table 5. SURVIVAL OF SYMBIOTIC ALGAE IN CHIMERAS OF *Stentor coeruleus* AND *Stentor polymorphus*

		Proportion of chimeras retaining algae*		
	coer./polym. volume ratio	Enucleated *S. coeruleus*	Partially enucleated *S. coeruleus*	Nucleated *S. coeruleus*
Nucleated	$>$ 4/1	8/19 (42%)	1/4 (25%)	6/35 (17%)
S. polymorphus	$\leq$ 4/1	9/11 (82%)	7/9 (78%)	19/35 (54%)

From Tartar, 1953.
* Chimeras which did not survive at least 3 days were excluded when they could be identified.
Classifications used in this table are necessarily arbitrary. The reader is urged to examine the original paper.

species which has no algal symbionts). The grafts were themselves of limited viability, although some regeneration occurred, and they usually persisted for a few days. Tartar's results show two things quite clearly: (1) The *coeruleus* cytoplasm is incompatible with the algae; the higher the proportion of *coeruleus* cytoplasm, the more likely the algae were to be lost. The basis of this incompatibility is not known. (2) The presence of a *coeruleus* nucleus enhances the probability that the algae will be lost. This appears to be the only direct demonstration of a nuclear role in the symbiosis, and these skillful experiments, which were not primarily designed for study of the symbiosis, deserve to be extended further.

The functional relationship between the organisms

The effect of the algae upon the growth of the host appears to be remarkably similar in *Paramecium bursaria* (Pringsheim, 1928; Karakashian, 1963; Pado, 1965, 1967), in *Stentor polymorphus* (Hämmerling, 1946; Schulze, 1951), and in *Chlorohydra viridissima* (Muscatine, 1961, 1965; Stiven, 1965). In each case the algae either significantly enhance the growth of the host in light when the exogenous food supply is limiting or prolong the survival of the host under starvation conditions (Table 6). (For a contrary report, see Park et al., 1967.) However, the algae have no effect on the growth rate when excess food is provided. In other words, the symbiosis appears to be functionally significant only when the food supply is depleted. In the case of *P. bursaria* at least, bacterial concentrations high enough to obscure the symbiotic enhancement of growth probably occur relatively rarely in nature, and so the symbiosis is almost certainly adaptive.

In both *Stentor* and *P. bursaria* the intracellular algae grow well in complete darkness, as long as the host is provided with food. This indicates that under these conditions the predominant direction of nutrient transfer is from host to algae. When starved in the dark, however, the host gradually digests its algae—a fact which may explain the slight growth stimulation observed in darkness. If starved sufficiently long, *Chlorella*-free lines can be obtained.

The most likely explanation of these and related data is that photosynthetic products from the algae are transferred in light to the host, where they are metabolized. Muscatine has provided direct evidence of this in two cases. In long-term labeling experiments, he and Hand (1958) demonstrated autoradiographically light-stimulated fixation of $C^{14}O_2$ by the zooxanthellae of the anemone *Anthopleura elegantissima* and the subsequent transfer of the label to the animal tissues. More recently, Muscatine and Lenhoff (1963) have shown a similar light-

Table 6. STIMULATION OF HOST GROWTH BY SYMBIOTIC ALGAE UNDER STARVATION CONDITIONS IN LIGHT

Chlorohydra viridissima (Muscatine, 1965)

	Mean no. of hydranths on day*							
	0	2	4	6	8	10	12	14
Infected	10	20.5	24.7	28.5	29.5	29.0	32.5	31.0
Uninfected	10	18.2	20.6	18.7	12.5	5.6	1.7	0.6

Paramecium bursaria (Karakashian, 1963)

	Mean no. of paramecia on day†								
	0	1	2	3	4	5	6	7	8
Infected	10	7.8	8.5	15	18	30	41	50	80
Uninfected	10	5.5	2.7	0	0	0	0	0	0

*Mean of 8 replicates.
†Mean of 6 replicates.

dependent fixation in *Chlorohydra viridissima,* with transfer of about 12 percent of the carbon fixed to the algae-free ectoderm of the host; chemical fractionation experiments showed that the labeled material was metabolized by the host cells. The presence of symbiotic algae also leads to a sharp reduction in protein turnover of the host (Muscatine and Lenhoff, 1965). The basis of this effect is unknown, but since partial complements of algae are nearly as effective as complete ones, and since infected heads are effective in slowing turnover in uninfected bodies to which they have been grafted, the agent may be a diffusible cofactor.

Although demonstrating clearly the light-stimulated transfer of metabolically active carbon, these experiments tell us nothing about the nature of the active compound or compounds. In fact, neither experiment rules out the possibility that digestion of intracellular algae may be the chief mechanism of carbon transfer. Fortunately some other data are available which bear on this question.

Muscatine and his coworkers (Muscatine, 1965; Muscatine et al., 1967) have performed a series of short-term $C^{14}O_2$ fixation experiments *in vitro* using both symbiotic and free-living *Chlorella.* Altogether 20 strains were examined, including 2 strains from *Chlorohydra viridissima,* 1 strain from the fresh water sponge, *Spongilla,* 7 strains from *P. bursaria,* and 10 free-living strains. Although the overall rates of photosynthesis were similar, the strains differed in important respects. All of the symbiotic strains, save the one from *Spongilla,* liberated a relatively high proportion of the carbon fixed into the medium

(15%-86%), while the proportion was much lower among the free-living strains (1%-8%). Moreover, the principal extracellular compound of the symbionts was a sugar, while among the free-living strains it was glycolic acid. All of the *P. bursaria* algae and one of the strains of hydra algae liberate maltose, while the sponge alga, and possibly the other strain of hydra algae, liberate glucose. (Representative data are presented in Table 7.)

Table 7. DISTRIBUTION OF C^{14} AFTER PHOTOSYNTHESIS
(30 min., 1,500 foot-candles)

Source of algae	Percent in medium	Percent as maltose	Percent as glucose	Percent as glycolic acid and unknowns
C. viridissima, CS60	56.4	84.1	12.1	3.7
Spongilla sp., 838	4.4	0	99.9	0.1
P. bursaria, NC64A	86.0	97.1	1.2	1.7
130C	15.3	97.0	0	3.0
34A	46.2	99.4	0.5	0.1
Free-living *Chlorella* GrB1	1.1	0	0	100
490	7.6	0	0	100
262	0.4	0	0	100
263C	0.5	0	0	100

From Muscatine, 1965, and Muscatine et al., 1967.

Although these experiments were performed *in vitro*, some limited data, summarized in Table 8, suggest that the extrapolation to the *in vivo* situation may be justified. Thirty cells of each of five subclones of a single *P. bursaria* stock were isolated individually into microcultures under semi-starvation conditions (2% baked lettuce infusion). One subclone was uninfected, one subclone was infected with a normal symbiotic alga (NC64A), and three subclones were each heavily infected with several hundred cells of a different free-living alga (262, 263C, and GrB1). After 3 days incubation in a diurnal regimen of 12 hours light and 12 hours darkness, the number of paramecia in each culture was recorded.

The effect of the algae on host growth parallels the liberation of extracellular compounds *in vitro*. Free-living strains, far from enhancing host growth, seem to have retarded it in this experiment, while the symbiotic algae have stimulated division. Although tracer experiments are needed to confirm these observations, the results do support the hypothesis that the transfer of soluble materials, rather than digestion, is responsible for the stimulation of host growth in light. There is no

Table 8. GROWTH OR SURVIVAL OF PARAMECIA INFECTED WITH FREE-LIVING CHLORELLA

Strain of algae	No. of paramecia per microculture after 3 days 1	2	3	4	>4
Uninfected	0	15	8	7	0
C. vulgaris, 262	9	18	2	1	0
C. vulgaris, 263C	10	18	2	0	0
C. vulgaris, GrB1	9	21	0	0	0
C. sp., NC64A (symbiotic)	0	0	4	22	4

The experiment was begun with one paramecium per microculture. Each sample consisted of 30 microcultures.

reason to suppose that free-living algae are less digestible than symbionts. This interpretation also receives support from the elegant work of Smith and his co-workers (Smith and Drew, 1965; Drew and Smith, 1966; Richardson and Smith, 1966; Richardson et al., 1967) who have shown that soluble sugars or their derivatives are translocated from the alga to the fungus in lichens—glucose, in the case of the lichenized blue-green alga, *Nostoc,* and the sugar alcohol ribitol, in the case of the green alga *Trebouxia.*

In summary, then, translocation of carbon from symbionts to host is probably mediated by the release of soluble carbon compounds in light, among them maltose and glucose, and these compounds are then metabolized by the hosts. Symbiotic algae, unlike free-living strains, appear to be adapted for the release of such compounds. Cytological observations indicate that digestion is probably rare in illuminated cultures, but during starvation in darkness the predominant mode of carbon transfer may be by digestion of algae. This latter process may well be related to autophagic vacuole formation which is known to be induced in starving cells, including protozoa (Levy and Elliott, 1968).

Speculations about the evolution of the Chlorella *symbioses*

A most fundamental observation is the striking similarity of the algae to one another and to free-living algae of the *Chlorella* type, while at the same time there is great diversity among the host organisms. These facts make it inconceivable that the various associations could have had a common ancestor, and a relatively recent polyphyletic origin is therefore indicated. The similarity of the algae points to a possible preadaptation for symbiosis of the genus *Chlorella* or related forms.

What evidence can be mustered in support of such an hypothesis? Evidence of preadaptation can be found in the relatively high proportion of free-living strains which are infective. These strains are able to persist for at least two days in nondividing paramecia—an interval 10 to 15 times longer than is required for digestion of food. The algae thus escape digestion by the host—clearly a prerequisite for the establishment of an infection. Unfortunately, the cellular events accompanying infections by even the normal symbionts are not yet understood. This problem is currently under investigation in our laboratory; when our study is completed we may then be able to discover what characteristics distinguish the infective and noninfective free-living strains.

What little evidence is available suggests that preadaptation has not occurred among species closely related to the host organisms. Thus there is no infection of other *Paramecium* species by *Chlorella* symbionts from *P. bursaria.* That the response of the hosts to symbiotic algae is conditioned at least in part by nuclear factors is suggested by Tartar's experiments on chimera-formation in *Stentor* (Table 5).

Lazo has neatly demonstrated that symbioses can arise between organisms which have apparently no previous symbiotic history. He was able to establish artificial associations between two species of myxomycetes and several strains of free-living *Chlorella.* The associations persisted indefinitely, and one of the slime molds was induced to fruit on an axenic medium which could not support the fruiting stage otherwise (Lazo, 1961). In a second series of experiments (Lazo, 1966; Lazo and Klein, 1965), a lichen-like thallus was induced in a mixed culture of a species of *Streptomyces* and *Chlorella xanthella.* The thallus, as in true lichens, contained discrete zones of algal and mycelial growth.

Once an initial association between the partners is formed, selection pressure should favor mutational changes leading to the sharing of a common carbon pool and to mutual adjustment of the growth rates. Both of these processes would increase the stability of the union. Undoubtedly, many free-living infections are abortive, but occasionally the variants necessary to achieve stability may arise soon enough to ensure the perpetuation of a permanently infected line.

Under such a polyphyletic scheme, variation from symbiosis to symbiosis ought to be the rule. One might predict, for example, that differing carbon compounds might be released by different algae, and indeed variation has been noted among the strains associated with different hosts (Table 7). Moreover, the ability to grow independently and the ability to infect other hosts would similarly be a function of the past history of each strain. On the other hand, the uniformity of the algae and those aspects of the intracellular environment common to all animal cells may impose some constraints on the variation and produce some convergence. For example, the presence of a single vacuole surrounding each alga maximizes the interface between the organisms and could be an adaptation for regulation of metabolic exchanges, in which the vacuole membrane may play a role.

The frequency with which new associations arise can only be a matter of conjecture. The variation among strains inhabiting a single species of host might well be a rough index of this frequency, but unfortunately the data on this point are not really adequate for an assessment.

ACKNOWLEDGMENT: Portions of the work described in this paper have been supported by National Science Foundation grants GB-3827, GB-6817, and GB-7362 to the author.

Literature Cited

Bomford, R. 1965. Infection of alga-free *Paramecium bursaria* with strains of *Chlorella, Scenedesmus,* and a yeast. J. Protozool. *12:* 221-224.

Bryan, G. W., A. H. Zadylak, and C. F. Ehret. 1967. Photoinduction of plastids and of chlorophyll in a *Chlorella* mutant. J. Cell Sci. *4:* 513-528.

Drew, E. A., and D. C. Smith. 1966. The physiology of the symbiosis in *Peltigera polydactyla* (Neck.) Hoffm. Lichenol. *3:* 197-201.

Droop, M. R. 1963. Algae and invertebrates in symbiosis. Sympos. Soc. Gen. Microbiol. *13:* 171-199.

Freudenthal, H. D. 1962. *Symbiodinium,* gen. nov., and *Symbiodinium microadriaticum,* sp. nov., a zooxanthella: taxonomy, life cycle, and morphology. J. Protozool. *9:* 45-52.

Hall, W. T., and G. Claus. 1963. Ultrastructural studies on the blue-green algal symbiont in *Cyanophora paradoxa* Korschikoff. J. Cell Biol. *19:* 551-563.

Hall, W. T., and G. Claus. 1967. Ultrastructural studies on the cyanelles of *Glaucocystis nostochinearum* Itzihsohn. J. Phycol. *3:* 37-51.

Hämmerling, J. 1946. Über die Symbiose von *Stentor polymorphus.* Biol. Zentr. *65:* 52-61.

Hood, C. L. 1927. The zoochlorellae of *Frontonia leucas.* Biol. Bull. *52:* 79-88.

Karakashian, S. J. 1963. Growth of *Paramecium bursaria* as influenced by the presence of algal symbionts. Physiol. Zool. *36:* 52-68.

Karakashian, S. J., and M. W. Karakashian. 1965. Evolution and symbiosis in the genus *Chlorella* and related algae. Evol. *19:* 368-377.

Karakashian, S. J., M. W. Karakashian, and M. A. Rudzinska. 1968. Electron microscopic observations on the symbiosis of *Paramecium bursaria* and its intracellular algae. J. Protozool. *15:* 113-128.

Lazo, W. R. 1961. Growth of green algae with Myxomycete plasmodia. Am. Midland Naturalist *65:* 381-383.

Lazo, W. R. 1966. An experimental association between *Chlorella xanthella* and a streptomyces. Am. J. Botan. *53:* 105-107.

Lazo, W. R., and R. M. Klein. 1965. Some physical factors involved in actinolichen formation. Mycologia *57:* 804-808.

Levy, M. R., and A. M. Elliott. 1968. Biochemical and ultrastructural changes in *Tetrahymena pyriformis* during starvation. J. Protozool. *15:* 208-222.

Loefer, J. B. 1936. Isolation and growth characteristics of the 'zoochlorella' of *Paramecium bursaria*. Am. Naturalist *70:* 184-187.

McLaughlin, J. J. A., and P. A. Zahl. 1959. Axenic zooxanthellae from various invertebrate hosts. Ann. N. Y. Acad. Sci. *77:* 55-72.

McLaughlin, J. J. A., and P. A. Zahl. 1966. Endozoic algae. In *Symbiosis,* Vol. 1, S. Mark Henry, editor. Academic Press.

Murakami, S., Y. Morimura, and A. Takamiya. 1963. Electron microscopic studies along (*sic*) cellular life cycle of *Chlorella ellipsoidea*. In *Studies on Microalgae and Photosynthetic Bacteria,* edited by the Japanese Society of Plant Physiologists. University of Tokyo Press.

Muscatine, L. M. 1961. Symbiosis in marine and fresh water coelenterates. In *The Biology of Hydra*. University of Miami Press.

Muscatine, L. M. 1965. Symbiosis of hydra and algae. III. Extracellular products of the algae. Comp. Biochem. Physiol. *16:* 77-92.

Muscatine, L. M., and C. Hand. 1958. Direct evidence for the transfer of materials from symbiotic algae to the tissues of a coelenterate. Proc. Natl. Acad. Sci. *44:* 1259-1263.

Muscatine, L. M., S. J. Karakashian, and M. W. Karakashian. 1967. Soluble extracellular products of algae symbiotic with a ciliate, a sponge, and a mutant hydra. Comp. Biochem. Physiol. *20:* 1-12.

Muscatine, L. M., and H. M. Lenhoff. 1963. Symbiosis: on the role of algae symbiotic with hydra. Science *142:* 956-958.

Muscatine, L. M., and H. M. Lenhoff. 1965. Symbiosis of hydra and algae. II. Effects of limited food and starvation on growth of symbiotic and aposymbiotic hydra. Biol. Bull. *129:* 316-328.

Northcote, D. H., K. J. Goulding, and R. W. Horne. 1958. The chemical composition and structure of the cell wall of *Chlorella pyrenoidosa*. Biochem. J. *70:* 391-397.

Oehler, R. 1922. Die Zellverbindung von *Paramecium bursaria* mit *Chlorella vulgaris* und anderen algen. Arb. Staatinst. Exptl. Therapie, Frankfurt a. M. *15:* 3-19.

Oschman, J. L. 1966. Development of the symbiosis of *Convoluta roscoffensis* Graff and *Platymonas* sp. J. Phycol. *2:* 105-111.

Oschman, J. L. 1967. Structure and reproduction of the algal symbionts of *Hydra viridis*. J. Phycol. *3:* 221-228.

Pado, R. 1965. Mutual relation of protozoans and symbiotic algae in *Paramecium bursaria*. I. The influence of light on the growth of the symbionts. Folia Biol. *13:* 173-182.

Pado, R. 1967. Mutual relation of protozoans and symbiotic algae in *Paramecium bursaria*. II. Photosynthesis. Acta Soc. Botan. Polon. *36:* 97-108.

Park, H. D., C. L. Greenblatt, C. F. T. Mattern, and C. R. Merril. 1967. Some relationships between *Chlorohydra,* its symbionts and some other chlorophyllous forms. J. Exp. Zool. *164:* 141-162.

Parke, M., and I. Manton. 1967. The specific identity of the algal symbiont in *Convoluta roscoffensis*. J. Mar. Biol. Assoc., U. K. *47:* 445-464.

Pringsheim, E. G. 1928. Physiologische Untersuchungen an *Paramecium bursaria*. Ein Beitrag zur Symbioseforschung. Arch. Protist. *64:* 289-418.

Richardson, D. H. S., and D. C. Smith. 1966. The physiology of the symbiosis in *Xanthoria aureola*. Lichenol. *3:* 202-206.

Richardson, D. H. S., D. C. Smith, and D. H. Lewis. 1967. Carbohydrate movement between the symbionts of lichens. Nature *214:* 879-882.

Ris, H., and W. Plaut. 1962. Ultrastructure of DNA-containing areas in the chloroplast of *Chlamydomonas*. J. Cell Biol. *13:* 383-391.

Schnepf, E., W. Koch, and G. Deichgräber. 1966. Zur Cytologie und taxonomischen Einordnung von *Glaucocystis*. Arch. Mikrobiol. *55:* 149-174.

Schulze, K. L. 1951. Experimentelle Untersuchungen über die Chlorellen-Symbiose bei Ciliaten. Biol. Generalis, Arch. Allgem. Frag. Lebensforsch. *19:* 281-298.

Shihira, I., and R. W. Kraus. 1963. *Chlorella. Physiology and Taxonomy of Forty-one Isolates.* Port City Press, Baltimore.

Siegel, R. W. 1960. Hereditary endosymbiosis in *Paramecium bursaria*. Exptl. Cell Res. *19:* 239-252.

Siegel, R. W., and S. J. Karakashian. 1959. Dissociation and restoration of endocellular symbiosis in *Paramecium bursaria*. Anat. Rec. *134:* 639.

Smith, D. C., and E. A. Drew. 1965. Studies in the physiology of lichens. V. Translocation from the algal layer to the medulla in *Peltigera polydactyla*. New Phytologist *64:* 195-200.

Soeder, C. J. 1964. Elektronenmikroskopische Untersuchungen an ungeteilten Zellen von *Chlorella fusca* Shihira et Kraus. Arch. Mikrobiol. *47:* 311-324.

Soeder, C. J. 1965. Elektronenmikroskopische Untersuchungen der Protoplastenteilung bei *Chlorella fusca* Shihira et Kraus. Arch. Mikrobiol. *50:* 368-377.

Staehelin, A. 1966. Die Ultrastruktur der Zellwand und des Chloroplasten von *Chlorella*. Z. Zellforsch. *74:* 325-350.

Starr, R. C. 1964. The culture collection of algae at Indiana University. Am. J. Botan. *51:* 1013-1044.

Stiven, A. E. 1965. The relationship between size, budding rate, and growth efficiency in three species of *Hydra*. Res. Pop. Ecol. *VII:* 1-15.

Tartar, V. 1953. Chimeras and nuclear transplantations in ciliates, *Stentor coeruleus* X *S. polymorphus*. J. Exptl. Zool. *124:* 63-103.

Taylor, D. L. 1968. Chloroplasts as symbiotic organelles in the digestive gland of *Elysia viridis* (Gastropoda: Opisthobranchia). J. Mar. Biol. Assoc., U. K. *48:* 1-15.

Vivier, E., A. Petitprez, and A. F. Chive. 1967. Observations ultrastructurales sur les Chlorelles symbiotes de *Paramecium bursaria*. Protistol. *III:* 325-334.

Biochemical Bridges Between Vascular Plants

PETER R. ATSATT
Department of Population and Environmental Biology
University of California, Irvine

VASCULAR PLANTS often do not grow as genetically distinct individuals, but are in fact physiologically interconnected with other individuals. Such biochemical or physiological bridges may be formed among the members of a single breeding population, or between the individuals of two or more species. The bridges themselves are formed by natural root or stem grafts, mycorrhizal fungi, or haustorial connections (Table 1). Graham and Bormann (1966) list more than 50 genera of angiosperms and over a dozen genera of gymnosperms that reportedly form natural root grafts. Pines and oaks are among the most common kinds of root-grafting trees. Natural stem grafts also occur, but they are less common.

Bridges formed by mycorrhizal fungi are particularly interesting. The flowering plant *Monotropa hypopitys* was originally regarded as a root parasite until its mycorrhizal fungi were discovered, and then it was labeled as a saprophyte. In 1960 Björkman demonstrated that *Monotropa* shares its mycorrhizal fungi with the roots of pine and spruce trees, and that nutrients pass from the trees to *Monotropa* via these fungal bridges. At least one member of the saprophytic Burmanniaceae family is also known to share its fungal associates with adjacent conifers (Table 1). Mycorrhizal sharing has been demonstrated between several nongreen orchids and such forest trees as *Fagus, Leptospermum,* and *Acacia* (Campbell, 1962, 1963, 1964). In addition, Ruinen (1953) has shown that many of the so-called epiphytic orchids are in fact connected to their host tree by fungal hyphae.

The fungal associations of forest trees have been amply documented but are not generally appreciated. Mycorrhizal fungi do not merely enhance but virtually enable the growth of a very large part of the world's commercially important plants (Wilde and Lafond, 1967). Wilde (1954) suggests that the occurrence of so-called "autotrophic" or nonmycorrhizal trees is a myth which has been thoroughly disproved. On the basis of radioactive tracer studies, Woods and Brock (1964) have suggested that mycorrhizal fungi may be mutually shared by the root systems of many forest trees. If future investigations confirm the concept of mycorrhizal sharing, then we may find that very few kinds of plants are immune from the direct biochemical influence of their neighbors.

Table 1. Genera and Families of Vascular Plants With Members That May Form Physiological Interconnections With Other Individuals Via Natural Grafts, Mycorrhizal Fungi, or Haustorial Connections

Natural Root[1] and/or Stem Grafts[2,3]			
Angiosperms	*Eugenia*[2]	*Pennantia*	*Ulmus*[2]
Acer	*Fagus*[2]	*Populus*	*Weinmannia*
Alectryon	*Ficus*[3]	*Posoqueria*	*Zelkova*
Alhagi	*Fraxinus*	*Prunus*	**Gymnosperms**
Arbutus	*Fuchsia*	*Pseudowintera*	*Abies*
Aristotelia	*Hedera*[2]	*Quercus*[2]	*Chamaecyparis*
Beilschmiedia	*Hevea*	*Rhus*	*Cryptomeria*
Betula	*Knightia*	*Robinia*	*Dacrydium*
Brachyglottis	*Laurelia*	*Salix*[2]	*Larix*
Carpinus	*Leucopogon*	*Sassafras*	*Libocedrus*
Carya	*Liquidambar*	*Schefflera*	*Picea*
Casuarina	*Manihot*	*Smilax*	*Pinus*[2]
Ceiba	*Metrosideros*	*Sophora*	*Pseudotsuga*
Cinnamomum	*Myrsine*	*Spathodea*	*Sequoia*
Citrus	*Myrtus*	*Suttonia*	*Taxodium*[2]
Coccoloba[2]	*Nothofagus*	*Symphoricarpos*	*Taxus*[2]
Corynocarpus	*Ochroma*	*Terminalia*	*Thuja*
Dysoxylum	*Olea*	*Tilia*	*Tsuga*
Eucalyptus			

Mycorrhizal Fungi		
Monotropaceae	Orchidaceae	Burmanniaceae
Monotropa hypopitys[4]	*Gastrodia* sp.[5]	*Thismia rodwayi*[6]

Haustorial Connections[7]			
Hemiparasites		**Holoparasites**	
Krameriaceae[8]	Olacaceae	Balanophoraceae	Loranthaceae
Lauraceae	Opiliaceae	Cuscutaceae	Orobanchaceae
Loranthaceae	Santalaceae	Cynomoriaceae	Podocarpaceae
Myzodendraceae	Scrophulariaceae	Hydnoraceae	Rafflesiaceae
		Lennoaceae	

[1] Graham and Bormann, 1966.
[2] Beddie, 1941.
[3] Rao, 1966.
[4] Björkman, 1960.
[5] Campbell, 1962, 1963, 1964.
[6] Campbell, 1968.
[7] Ozenda, 1965.
[8] Cannon, 1910.

Haustorial connections are found in plant families with members that are either partially parasitic (hemiparasites) or completely parasitic (holoparasites). Interplant connections in this group are formed by haustoria, which are specialized penetrating tissues initiated by the parasite. They are usually identifiable as swollen, knob-like structures at the points of contact with other roots.

Unfortunately, we understand very little about the specific kinds of compounds that are normally translocated between interconnected individuals, especially the form in which these compounds are donated and the manner in which they are utilized. Several investigators have demonstrated translocation between grafted trees by using dyes, radioactive isotopes, silvicides, fungus spores, and antibiotics (Graham and Bormann, 1966). There is also some evidence that auxin transport through grafts may influence the initiation of cambial activity in recipient trees (Wold and Lanner, 1965). It has been demonstrated that plant viruses are transmitted from one plant to another via the haustorial connections of several species of dodder (Sakimura, 1947; Kunkel, 1952). Transport of labeled carbon (initially incorporated in CO_2 or sugars) and labeled minerals has been demonstrated across both mycorrhizal bridges (Björkman, 1960) and the haustorial connections of numerous hemiparasites (Rogers and Nelson, 1962; Okonkwo, 1966; Govier, Nelson, and Pate, 1966; and others).

One of the most promising lines of inquiry into the nutrition of hemiparasites is that of Govier, Nelson, and Pate (1966). Following a short exposure of the host to $C^{14}O_2$, the authors were able to recover labeled amides, amino acids, organic acids, and sugars in the sap exuding from cut root stumps of the host and from the attached hemiparasite *Odontites verna*. Their results suggest that photosynthetically fixed carbon is translocated to the host roots as sucrose and then rapidly converted to other compounds within the haustorial tissues of the hemiparasite. Approximately 53 percent of the label in the hemiparasite sap was recovered as glucose and fructose, 12 percent in compounds thought to be organic acids, and the remaining 35 percent was attached to amino acids and amides.

From the biochemist's point of view, the demonstration that both organic and inorganic compounds travel between interconnected individuals is certainly inadequate. However, population biologists now possess a nucleus of information that will allow a beginning approach to questions of the genetic and evolutionary significance of biochemical exchange between genetically distinct individuals. But we must have a set of experimental tools, a set of organisms that can be manipulated both in the laboratory and in the field. Hemiparasites, particularly annual hemiparasites, prove to be excellent experimental material.

Some Characteristics of Hemiparasites

Two characteristics are implicit in the term hemiparasite: the presence of haustoria and the presence of functional chlorophyll. The prefix "hemi" suggests that these plants are partially autotrophic and partially parasitic, but the degree of dependency on a host is not quantified. Actually, the presence of chlorophyll is of little value in determining the degree to which one plant is dependent upon another. The common mistletoe, *Phoradendron,* is an extreme example. It has been demonstrated recently (Hull and Leonard, 1964) that some *Phoradendron* species supply most of their own carbohydrates, yet they are obligate parasites because they lack functional roots; they require a living host to complete their life cycle. The commercially important sandalwood tree from India and the Hawaiian Islands is also a hemiparasite. It is photosynthetic and has a well-developed root system, but it will not grow and reproduce without the aid of one or several host plants (Rao, 1911).

Hemiparasites grow in a variety of habitats, ranging from equatorial to subpolar latitudes. The more than 2,000 species include herbaceous annuals, biennials, perennials, woody shrubs, vines, and trees. There seems to be very little correlation between hemiparasitism and any particular kind of environmental stress. In fact, these very different life forms occupy such distinctive habitats and graft to such different kinds of hosts that the physiological significance of grafting is likely to be quite different in each case. However, if root grafting selectively influences survival and reproduction, then it will also influence genetic recombination, and we see immediately that root grafting may be a major component of the genetic system of a population. The total effect, however, will depend upon how root grafting is integrated with all the factors that affect recombination, in particular the breeding system of a species.

Another interesting feature of hemiparasite populations is the apparently indiscriminate pattern of grafting. The members of most populations have the ability to graft to most of the individuals that surround them, so that in populations with high seedling density, the majority of early grafts will be between members of the same breeding population. If the root systems of several species are available, a single hemiparasite may graft to more than one individual, either at one time or during some portion of its life cycle.

Annual Hemiparasites

Annual hemiparasites exhibit considerable variation in growth with and without a host. Figure 1 illustrates the growth of individuals from four different genera of the family Scrophulariaceae. *Cordylanthus filifolius* blooms in the summer and the other three annuals bloom in the spring. *Cordylanthus* grows in open grassy areas in the chaparral, while *Castilleja stenantha* is found in naturally disturbed areas along stream banks. *Orthocarpus purpurascens* and *Parentucellia viscosa* are both grassland annuals and often occur in disturbed pastures.

Samples of the four species were raised together in a single growth chamber under uniform conditions (50° night, 70°, 12-hour day). Each line represents the average height obtained by three individuals when grown singly (dashed line), in clusters of three (broken line), and singly, but with the host plant, *Hypochoeris glabra* (solid line), a member of the family Compositae. For three species the growth pattern with a host is quite similar, with extremely rapid growth beginning between the second and fourth week after germination. It is quite apparent that *Hypochoeris glabra* is not a suitable host for *Parentucellia viscosa;* however, the presence of a host did keep these individuals alive, whereas seedlings without a host died during the fifth week. Single *Cordylanthus* plants also grew poorly as did the single *Castilleja* plants, although these survived the full 12 weeks. Single *Orthocarpus purpurascens* individuals, on the other hand, grew almost as well without hosts as with host plants. Thus, we have some index of the genetic capacity these individuals have for independent growth and also of the effect of providing a host plant. Now consider the individuals grown without host plants but in clusters of three. Clustering increased the rate of growth in *Cordylanthus, Parentucellia,* and *Castilleja* and prolonged the life span of *Cordylanthus* seedlings. The growth response of clustered *Orthocarpus* individuals suggests that some sort of competition occurred among these plants, but later examples will demonstrate that this is not always true. Intraspecific grafting has been observed in all four species, indicating that the differences in growth and survival may be due to some physiological interaction among the interconnected members of a cluster.

These data suggest that annual hemiparasites have several very different potentials, one determined by their own genotype, another related to intraspecific grafting, and a third potential governed by interaction with the host plant.

The selective influence of the host environment cannot be understood until we understand the reciprocal interactions that occur between the members of a single breeding population. Even at this level, the interaction will remain unintelligible until the relationship between specific genotypes and the resultant phenotypic expression can be defined. As a member of a hemiparasite population, is it "who you are" or "who you are with" that determines your destiny? Fortunately, this question can be answered because the members of many hemiparasite populations can be grown singly, without hosts, in the laboratory. By eliminating the host influence, the genetic potential of both individuals and of population samples can be tested.

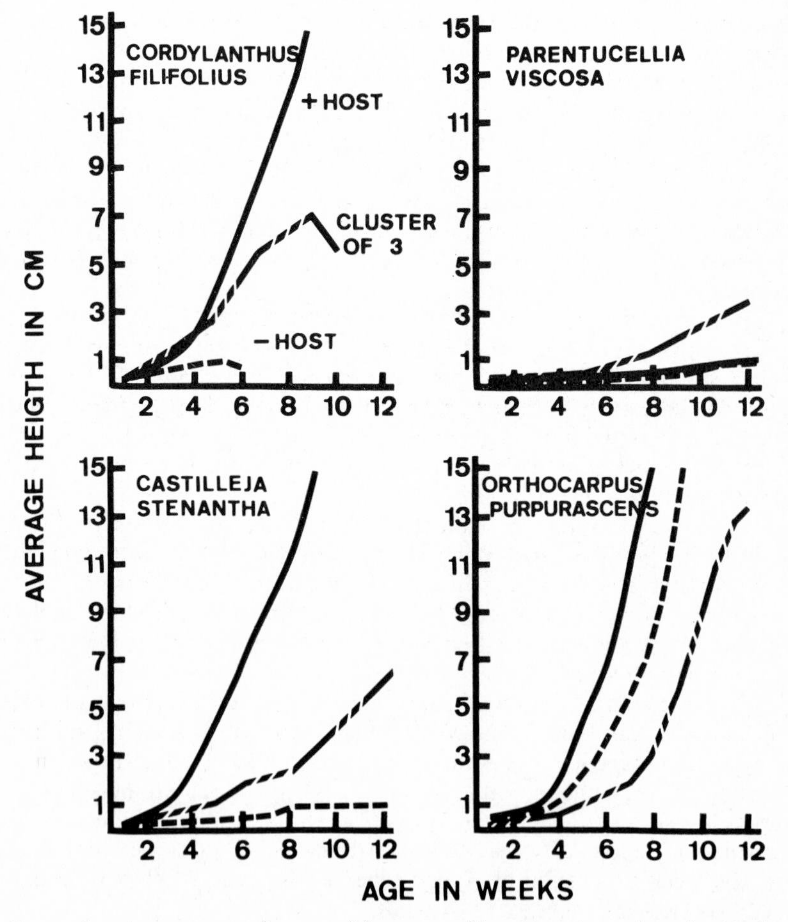

FIGURE 1. Average growth rates of four annual hemiparasites cultured with the host *Hypochoeris glabra* (solid line) and without a host, both singly (dashed line) and in clusters of three (broken line).

Genetic Variability

Orthocarpus purpurascens was selected for experimentation because its phenotypic variability suggested inherent genetic variability. The phenotypic polymorphism of these populations is in part due to their breeding system; the individuals are self-incompatible and are obligately outcrossed by a variety of insects. Breeding systems in *Orthocarpus* tend to function at one extreme or the other, promoting either obligate outcrossing or automatic self-pollination. The genetic systems of the inbreeders are significantly different from those of the outbreeders, and preliminary investigations indicate that we cannot generalize from one system to the other.

Most population samples of *O. purpurascens* exhibit considerable variation in their capacity for independent growth. Figure 2a illustrates the maximum heights attained by 150 individuals of an *O. purpurascens* population grown without hosts in 4-inch plastic pots containing a mixture of sand and peat moss. The pots were buried in a similar soil mixture to provide a uniform root environment and the plants were grown in an outdoor screenhouse from January through May.

Individuals less than 10 cm in height normally do not flower, so that in this sample (generation 1) 30 percent of the plants died before reaching sexual maturity. In other tests over 50 percent of the individuals died prior to flowering. Evidently, the members of these populations are genetically variable in their capacity to produce a complete metabolic quota, and in the absence of a host as many as half of these deficiencies may be lethal.

There are then two classes of individuals in these populations: those having the genetic capacity to grow and reproduce without the benefit of a host (10 cm and above) and those that absolutely require a host to mature and reproduce. We now need a hypothesis to explain how two such genetically distinct sets of individuals can be maintained in the same breeding population.

The presence of lethal and sublethal genetic deficiencies in these populations can best be explained by the concept of physiological buffering, or sharing of gene products. Individuals lacking an essential compound may obtain that compound from genetically complementary individuals. Thus, many genotypes that would normally be nonreproductive are apparently maintained in these populations and contribute to the genetic variation of future generations. The net effect is relaxed selection against aberrant individuals.

In order to account for the presence of the more independent individuals, there must be continuous or intermittent selection for the

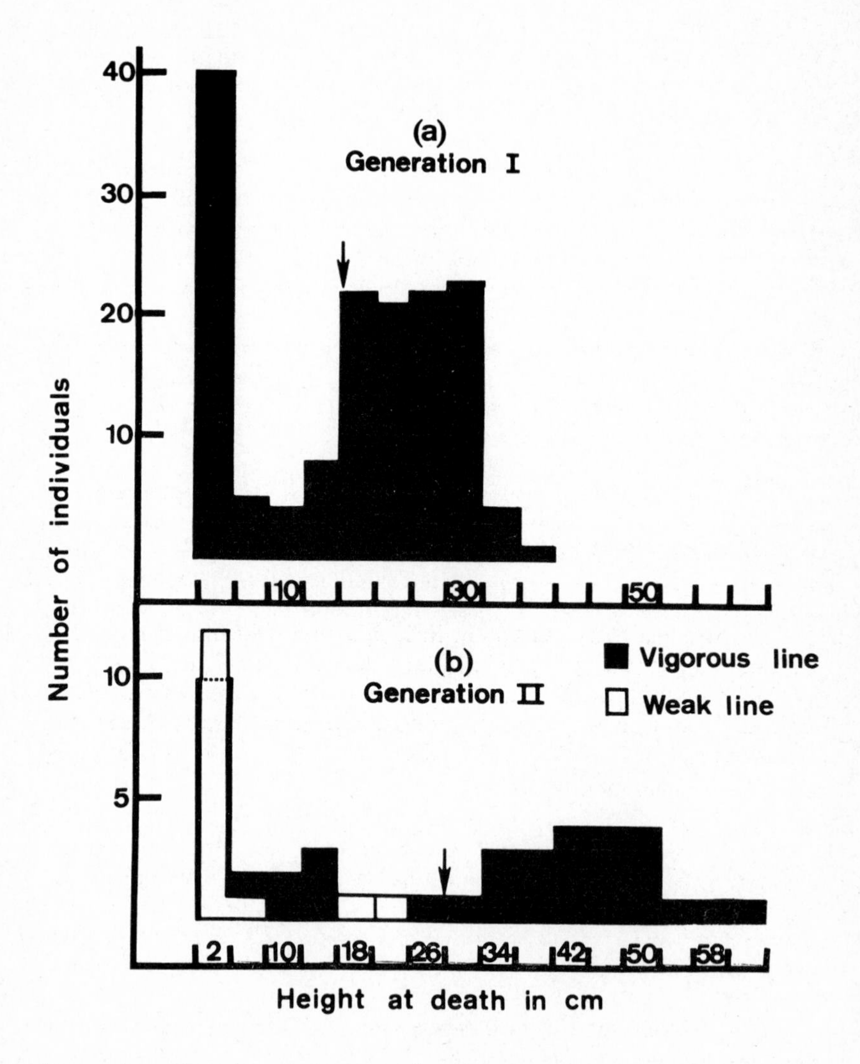

FIGURE 2a. Generation I: Maximum heights attained by 150 individuals of an *Orthocarpus purpurascens* population grown without hosts in 4-inch plastic pots. The abscissa values represent the median of each size class. Individuals less than 10 cm tall did not reproduce.

FIGURE 2b. Generation II: Maximum heights attained in two lines of selection initiated by reciprocal cross-pollination of the two largest individuals (vigorous line) and of the two smallest individuals (weak line) from generation I.

ability to grow independently. Selection for autotrophism seems to occur several ways in natural populations, but for the moment, let us discuss artificial selection for independent growth.

Two lines of selection in plants without hosts were initiated by reciprocal cross-pollination of the two largest individuals (vigorous line) and of the two smallest individuals (weak line) from generation I (Figure 2a). Poor germination in the second generation limited the number of individuals in the weak line to 16, while the vigorous line consisted of 40 individuals. The plants were grown in 4-inch plastic pots in an environmental chamber with a regime of 50° nights and 70°, 12-hour days. Unfortunately, these conditions differed slightly from those of the first generation, so that direct comparison of the two generations is subject to some interpretation.

The distribution of heights at death in the second generation is shown in Figure 2b. Individuals 14 cm or less in height died as seedlings; however, 15 out of these 16 deaths can be attributed to localized adverse wind conditions within the chamber. The 15 pots were in a single pan that received a constant wind deflected by an exhaust fan, which markedly increased evaporation from both the soil surface and the base pan. There is every reason to believe that if the 15 seedlings had not been subjected to this constant breeze, they would have performed as well as the rest of the progeny sample. This suggests that water balance is a very critical factor influencing both survival and growth of single individuals.

In the weak line, all but two individuals died before flowering, illustrating that segregation for vigorous autotrophs does not occur when weak individuals interbreed. Cross-breeding two vigorous individuals produced a range of healthy segregants and increased average height by 12 cm. Maximum height was increased 22 cm, or 55 percent.

The response of this population sample to artificial selection suggests that the genetic control of autotrophism in natural populations is delicately balanced, and that vigorous independent growth may be determined by a large number of genes with additive effects. The primary goal in establishing two selected lines has not yet been achieved. In the third and later generations, sub-samples of both the vigorous line and the weak line will be grown with host plants. It will then be possible to demonstrate the phenotypic effects of a host in the two opposing genetic strains, and, reciprocally, the effect that these known genotypes have on the host species.

Intraspecific Grafting

Hemiparasites often grow better in clusters than they do as isolated plants (Cantlon, Curtis, and Malcolm, 1963; Yeo, 1964). Variable results are obtained when *Orthocarpus purpurascens* individuals are grown in clusters of three. All three plants may die, all three may grow at an even rate, one may grow rapidly and the other two slowly, or any of the other possible growth combinations may occur. Mutual benefit seems to occur in some cases, while in others the interaction appears to be deleterious to one or more members of the cluster. A significant pattern does emerge, however, when a reasonably large population sample is studied. Seed from a single population of *Orthocarpus purpurascens* was subdivided and one set of 50 individuals was grown singly in 4-inch plastic pots and another group of 60 individuals was grown in clusters of three plants per pot. The most striking result of this experiment was the differential survival in the two treatments. Only 48 percent of the plants grown singly survived and flowered, while growth in clusters of three raised survival to 72 percent, a remarkable 24 percent increase. The distribution of dry weights in each treatment is given in Figure 3. The high percentage of individuals in the lowest weight class is a consequence of the early deaths of single individuals. Single individuals tended to be either large or very small, while clusters of three produced a higher percentage of intermediate plants and fewer extremes. In this case, intraspecific grafting had a mediating or averaging influence. The members of a cluster apparently receive something from grafted partners, allowing the survival of individuals that would normally perish. A possible explanation is that the physiological interaction of two or more genetically complementary individuals may overcome otherwise lethal deficiencies. This sort of physiological buffering is only plausible among highly heterozygous populations, of which *Orthocarpus purpurascens* is an example. It is most reasonable that an *Orthocarpus* deficiency could best be corrected by the gene products of another genetically complementary *Orthocarpus* individual. Intraspecific sharing, however, does not substitute for the energy source normally received from other kinds of plants.

Another suggestion is that the formation of intraspecific grafts enhances root uptake capacity (Heinricher, 1909). This sort of mechanism could operate in the following way: Members of *Orthocarpus* populations have differing capacities for independent growth, and the most vigorous in a cluster will be the largest at the time intraspecific grafts are formed. Under conditions conducive to moderately high transpiration, those individuals with the greatest leaf area will command a major portion of the materials flowing in the xylem. These

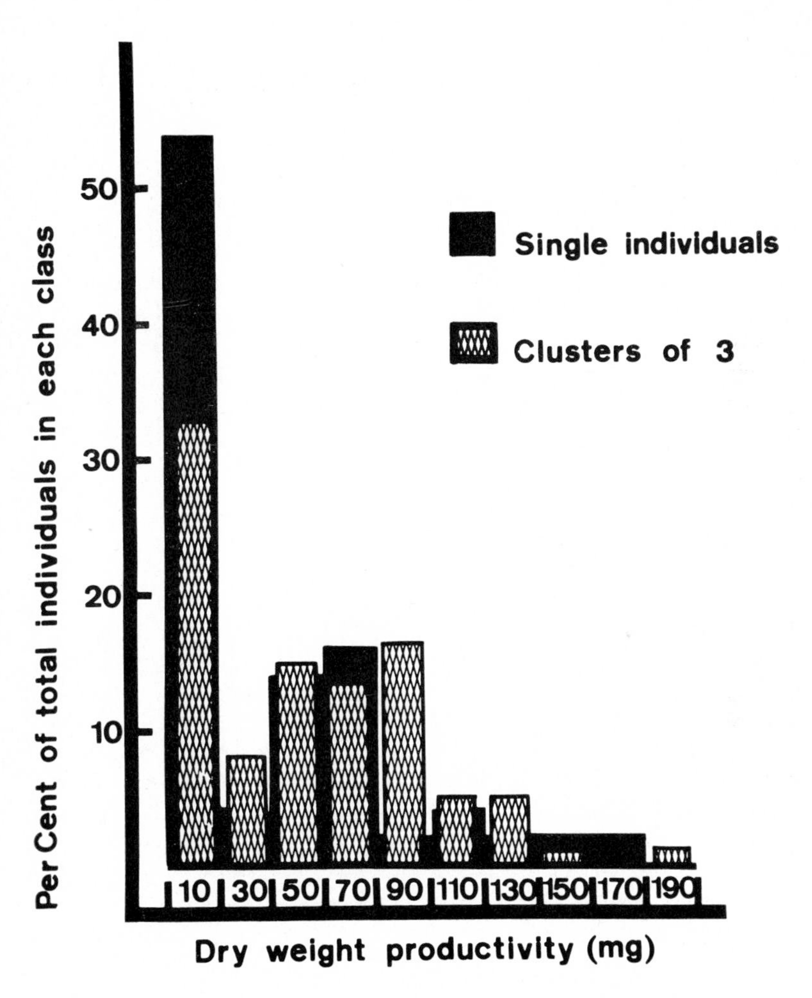

FIGURE 3. Distribution of the dry weights of *Orthocarpus purpurascens* individuals grown singly and in clusters of three. The abscissa values represent the median of each weight class.

larger individuals may then provide the necessary transpirational "pull" to increase the uptake capacity of smaller grafted neighbors, thereby maintaining a favorable water balance within the cluster. Continuing rapid growth of the dominant individual can, however, lead to the demise of the smaller plants in the cluster.

Within a population of hemiparasite seedlings, there is both competitive interaction and, at the same time, some degree of mutual benefit. Bormann (1966) has independently reached similar conclusions about intraspecific root grafting in *Pinus strobus.* His concluding paragraph states:

> This study suggests that the development of a naturally occurring white pine stand is shaped by two contrasting ecological forces: (1) Competition and (2) a noncompetitive force governed by intertree food translocation. Competition operates to reduce the number of individuals and to increase the volume of environment occupied by each individual. The noncompetitive force, on the other hand, counteracts the effects of competition by delaying the death of individuals. Competition is the more important force.

Clustering is a consistent feature among both root-grafting and haustoria-forming plants. There can be little doubt that intraspecific grafting confers advantages on a majority of the members of a cluster. Such a system could not have evolved under natural selection unless the individuals associated in groups left more offspring than did isolated individuals. The next question to be answered concerns the effect that intraspecific sharing may have on the kinds and frequency of genotypes that survive each generation.

The Host-Environment

As *Orthocarpus* seedlings mature, the probability increases that their root systems will come into contact with the roots of other species. Unlike pine trees, these plants now begin to form bridges with a diverse array of genetically different individuals, including members of the grass, legume, and sunflower families (Figure 4). Each of these hosts has special physical, temporal, and chemical characteristics which in combination produce a very permissive environment. By tapping the diverse biochemical products of neighboring plants, hemiparasites gain considerable latitude for genetic experimentation. As a consequence of the relaxation of selection allowed by physiological buffering, inferior alleles may be retained in the population because their effects are not disadvantageous in the host-environment. It is reasonable to assume that these populations would begin to accumulate genetic deficiencies and gradually evolve towards increasing host dependency. Yet, I have suggested that this is not the case, that some populations maintain a genetic equilibrium that allows a portion of their members to function both autotrophically and heterotrophically. This nutritional balance can be maintained only by continuous selection for autotrophism.

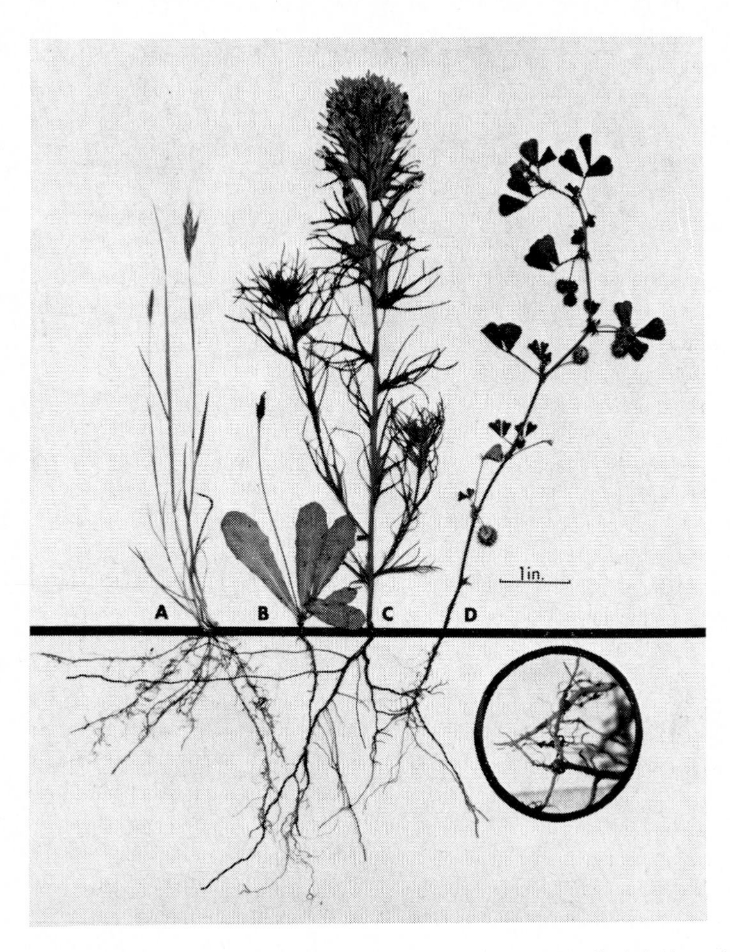

FIGURE 4. Schematic reconstruction of an *Orthocarpus* individual (C) grafted to the roots of three different hosts: *Festuca megalura* (A), *Hypochoeris glabra* (B), and *Medicago hispida* (D). (From a grassland-pasture near Goleta, California.)

Natural selection for autotrophic genotypes may operate both before and after host contact. Nearly all hemiparasites grow independently during early seedling growth and must be self-sustaining until grafts are formed. Intraspecific grafting then provides a powerful sorting mechanism, since similar but distinct genotypes are pitted one against the other. Intraspecific competition may be the mechanism that regulates the kinds of genotypes that are eventually exposed to the host-environment. The most vigorous competitors in a cluster are likely to make the most host contacts and continue to grow at a faster rate, ultimately producing the most offspring. Less vigorous seedlings may be killed before they become attached to a host. Of course, many individuals in the population are isolated from each other and graft directly with one or more host plants.

Any buildup of deleterious genes is likely to be significantly reduced during seasons of drought. Decreased moisture leads to a reduction in the number and kinds of plants available as hosts, their root systems are smaller, and many potential hosts are so impoverished that they are of little benefit to the hemiparasite population. Under these conditions, genetically deficient individuals will be quickly eliminated, and a premium will be placed on the capacity for autotrophic growth.

Continuous selection for autotrophism provides some genetic cohesion in a system which might otherwise become increasingly unstable. Both *Orthocarpus* populations and their myriads of annual hosts must reestablish themselves each year from seed. It is very unlikely that the progeny of successful hemiparasites will come into contact with the same host environment in which their parents prospered. This heterogeneous, fluctuating environment demands an equally versatile hemiparasite gene pool, and apparently most outbreeding populations do maintain a high level of genetic variability. Introgressive hybridization is quite common in *Orthocarpus,* (Thurman, 1965; Atsatt, 1966) and morphological variability is usually pronounced, both within and between local breeding populations. A wide range of recombinant genotypes is apparently maintained by physiological buffering from host plants, but this disruptive tendency is balanced by the stabilizing effect of continued selection for autotrophism.

Some hemiparasites really do deserve their name; they are partially autotrophic and partially parasitic, and it is this balance between two modes of existence that provides them with both immediate fitness and genetic flexibility for long-term survival. The perpetuation of this genetic flexibility is the insurance that these populations carry—insurance that they will be able to function effectively with a majority of their potential hosts. Loss of this variability would undoubtedly lead to increasing host specialization and decreased potential for adaptive radiation.

Acknowledgment: The research on which this paper is based was supported by National Science Foundation grant GB-6358.

Literature Cited

Atsatt, P. R. 1966. The population biology of closely related species of *Orthocarpus*. Ph.D. dissertation, University of California, Los Angeles.

Beddie, A. D. 1941. Natural root grafts in New Zealand trees. Trans. Roy. Soc. N. Z. *71:* 199-203.

Björkman, Erik. 1960. *Monotropa hypopitys* L.—an epiparasite on tree roots. Physiol. Plant. *13:* 308-327.

Bormann, F. H. 1966. The structure, function, and ecological significance of root grafts in *Pinus strobus* L. Ecol. Monographs *36:* 1-26.

Campbell, E. O. 1962. The mycorrhiza of *Gastrodia cunninghamii* Hook. f. Trans. Roy. Soc. N. Z. *1:* 289-296.

Campbell, E. O. 1963. *Gastrodia minor* Petrie, an epiparasite of Manuka. Trans. Roy. Soc. N. Z. *2:* 73-81.

Campbell, E. O. 1964. The fungal association in a colony of *Gastrodia sesamoides* R. Br. Trans. Roy. Soc. N. Z. *2:* 237-246.

Campbell, E. O. 1968. Personal communication.

Cannon, W. A. 1910. Parasitism of *Krameria canescens*. Carnegie Inst. of Washington, No. 129, 21 pp.

Cantlon, J. E., E. J. C. Curtis, and W. M. Malcolm. 1963. Studies of *Melampyrum lineare*. Ecology *44:* 466-474.

Govier, R. N., M. D. Nelson, and J. S. Pate. 1966. Hemiparasitic nutrition in angiosperms. I. The transfer of organic compounds from host to *Odontites verna* (Bell.) Dum. (Scrophulariaceae). New Phytologist *66:* 285-297.

Graham, B. F. Jr., and F. H. Bormann. 1966. Natural root grafts. Botan. Rev. *32:* 255-292.

Harley, J. L. 1959. In *The Biology of Mycorrhiza*. London: Leonard Hill Limited.

Heinricher, E. 1909. Die grunen Halbschmarotzer. V. *Melampyrum*. Jahrb. Wiss. Botan. *46:* 273-376.

Hull, R. J., and O. A. Leonard. 1964. Physiological aspects of parasitism in mistletoes (*Arceuthobium* and *Phoradendron*) I. The carbohydrate nutrition of mistletoes. Plant Physiol. *39:* 996-1007.

Kunkel, L. O. 1952. Transmission of alfalfa witches' broom to non-leguminous plants by dodder and cure in periwinkle by heat. Phytopathology *42:* 27-31.

Okonkwo, S. N. C. 1966. Studies on *Striga senegalensis* II. Translocation of C^{14}-labelled photosynthate, urea-C^{14} and sulphur-35 between host and parasite. Am. J. Botan. *53:* 142-148.

Ozenda, P. 1965. Recherches sur les phanerogames parasites. I. Revue des travaux recents. Phytomorphology *15:* 311-338.

Rao, A. N. 1966. Developmental anatomy of natural root grafts in *Ficus globosa*. Austral. J. Botan. *14:* 269-276.

Rao, M. 1911. The host plants of the Sandal Tree. Indian Forest Records *2:* 159-207.

Rogers, W. E., and R. R. Nelson. 1962. Penetration and nutrition of *Striga asiatica*. Phytopathology *52:* 1064-1070.

Ruinen, J. 1953. Epiphytosis. A second view of epiphytism. Ann. Bogor *1:* 101-157.

Sakimura, K. 1947. Virus transmission by *Cuscuta sandwichiana.* Phytopathology *37:* 66-67.

Thurman, L. D. 1965. Genecological studies in *Orthocarpus,* subgenus *Triphysaria* (Scrophulariaceae). Ph.D. dissertation, University of California, Berkeley.

Wilde, S. A. 1954. Mycorrhizal fungi: Their distribution and effect on tree growth. Soil Sci. *78:* 23-31.

Wilde, S. A., and A. Lafond. 1967. Symbiotrophy of Lignophytes and fungi: its terminological and conceptual deficiencies. Botan. Rev. *33:* 99-104.

Wold, M. L., and R. M. Lanner. 1965. New stool shoots from a 20-year-old swamp-mahogany *Eucalyptus* stump. Ecology *46:* 755-756.

Woods, F. W., and K. Brock. 1964. Interspecific transfer of Ca^{45} and P^{32} by root systems. Ecology *45:* 886-889.

Yeo, P. F. 1964. The growth of *Euphrasia* in cultivation. Watsonia *6:* 1-24.

Plant Poisons in a Terrestrial Food Chain and Implications for Mimicry Theory

Lincoln Pierson Brower
Department of Biology
Amherst College

The cycling of chemical substances through terrestrial and fresh-water ecosystems has attracted much attention in recent years as a result of the ever-increasing use of pesticides. DDT and other chlorinated hydrocarbons are well known for their accumulation in the food web of *Homo sapiens* (Duggan and Weatherwax, 1967). Notwithstanding this recent dramatic example, the idea that naturally occurring poisons are transferred from plants to herbivores and thence to predators is nearly a century old. Thus, Slater (1877) was impressed by the fact that many groups of warningly colored insects feed upon taxa of plants known to be poisonous, and he suggested that the insects derived their unpalatability by incorporating the noxious substances.

Clearly, this is a most important theory. More precisely, (1) plants are held to have evolved the ability to synthesize noxious chemical compounds which act as herbivore deterrents (it is well to remember that plants are attacked by numerous invertebrates and vertebrates below as well as above the ground) ; (2) certain groups of herbivores have not only evolved mechanisms to circumvent the toxic effects of the poisons, but (3) they selectively incorporate and in turn utilize the poisons to deter their own predators. In other words, at least three links in the food chain are thought to be involved in a complex ecological interaction.

Many correlative arguments, indirect lines of evidence, and uncontrolled feeding experiments have been produced over the years to support this theory (see Brower and Brower, 1964; Ehrlich and Raven, 1965, 1967; Eisner and Meinwald, 1966, for recent reviews). Since 1963, a concerted effort has been made in England, Switzerland, and our laboratories at Amherst College and in Trinidad, W.I., to carry out a thorough experimental verification. The approach has involved pharmacological assays, chemical identifications, and feeding experiments with caged predators under carefully controlled conditions.

The Organisms

The organisms we have studied are milkweed plants of the family Asclepiadaceae, butterflies of the subfamily Danainae which are almost completely restricted to the milkweeds as larval foodplants, and avian predators (particularly blue jays, *Cyanocitta cristata bromia* Oberholser, family Corvidae) which include insects in their diet in the wild state. These appeared to be good bioassay materials. Asclepiad plants are pharmacologically well known because many of them contain large amounts of cardiac glycosides (Hoch, 1961; Abisch and Reichstein, 1962) which are chemically similar to digitalis and are potent vertebrate heart poisons. The general effect of these substances on vertebrates is to decrease the frequency and increase the amplitude of the heartbeat; a large dose via the intravenous route causes heart failure and death, but if given orally it produces nausea and emesis, thus preventing death (Gero, 1965; Moe and Farah, 1965). One Danaine butterfly, the Monarch *Danaus plexippus* (L.), feeds on a wide variety of Asclepiad plants (Urquhart, 1960; Brower, 1961) and has long been recognized as the classical North American example of an unpalatable warningly colored insect. Experimental verification of the unacceptability of this butterfly as compared to controls has been shown for four species of birds belonging to two different families (J. Brower, 1958; Brower and Brower, 1964, and unpublished results), though several reports of wild birds eating Monarchs have appeared over the years, including the startling (yet, as we shall later see, completely explicable) results of Petersen (1964).

The Question

The specific hypothesis to be considered in this contribution is: do Monarch butterflies incorporate the cardiac glycosides from their milkweed foodplants and thereby become unacceptable (i.e., "unpalatable") to vertebrate predators, in particular the blue jay?

Pharmacological and Chemical Studies

The tropical American milkweed, *Asclepias curassavica* L., is a rich source of cardiac glycosides and is the principal foodplant of Monarch butterflies in Trinidad, W.I. (Brower, personal observations). Seeds from Mayaro, Trinidad, were taken to our Amherst College greenhouse where we have cultivated the plant and maintained breeding stocks of Monarchs with ease. Dried leaves of over 100 of these plants (there was a remote possibility of inclusion of three other *A. curassavica* plants of unknown origin which had been planted in our greenhouse from Mount Holyoke College) were analyzed by Santavy, Euw, and Reichstein (see Reichstein, 1967, reference no. 82) and found to contain large amounts of five cardiac glycosides (uscharin, uscharidin, calactin, calotropin, and calotoxin) as well as others not yet identified.

While in Trinidad during the summer of 1963, we reared approximately 100 chrysalids of Monarch butterflies on this plant and air mailed them to England for pharmacological investigations. Parsons (1965) homogenized the freshly killed adult and pupal material, extracted for cardiac glycosides, and separated three different substances by chromatography which he termed plexippin A, B, and C. Pharmacological tests of the extracted material were made using the various assays for digitalis, including guinea-pig ileum and suspended frog heart infusions, lethal dose determinations for cats via the intravenous route, and emetic dosages for starlings and pigeons via the oral route. Comparison with the standard digitalis group of controls strongly supported the contention that the butterflies do contain cardiac glycosides.

Subsequently, at Amherst College we reared approximately one kg of Monarchs (from a new stock obtained in Highlands County, Florida) on the same plants, preserved them as adults in alcohol, and shipped them to Professor Reichstein in Switzerland who analyzed them chemically (Reichstein, 1967). His investigation showed that these animals contained large quantities of cardiac glycosides. The two main components were calactin and calotropin. These were isolated as crystals and clearly identified. Eight others were also present in small quantities, including calotoxin and calotropagenin. The yield per butterfly of calactin, calotropin, and calotoxin was .04 to .05 mg. This contrasted with Parsons' (1965) larger estimate of .2 mg per butterfly and probably resulted from auto-oxidation of about 80 percent of the cardiac glycosides in the alcohol preservative. Reichstein (in manuscript) has also been able to show that Parsons' plexippin A and B correspond to calactin and calotropin, while C is a mixture of calotropagenin and calotoxin.

The difference in amounts of the various components in the plant and the butterfly suggests either that selective assimilation takes place, or that chemical conversion of the plant glycosides occurs in the animal. Reichstein (1967) and co-workers (*in* Euw, Fishelson, Parsons, Reichstein, and Rothschild, 1967) have carried out a parallel study on the grasshopper, *Poekilocerus bufonius* (Klug), fed upon *Asclepias curassavica* and another Asclepiad, *Calotropis procera* L., with comparable results.

Thus, pharmacological investigations and chemical analyses of milkweeds and two widely different insects which eat these plants provide strong support for the theory that some insects do have the ability to obtain their noxious properties from their foodplants. This is clearly the simplest explanation of the chemical identity of the various cardiac glycosides in the herbivores and their foodplants. Moreover, the fact that no insects have yet been discovered which can synthesize steroids from simple precursors (see Reichstein, 1967, for review) makes the case even stronger. Nevertheless, as we have pointed out (Brower, Brower, and Corvino, 1967) final disproof of the alternative of synthesis of cardiac glycosides by these particular insects (and it must be admitted that both *Danaus* and *Poekilocerus* have highly developed defensive systems) can only be achieved by radioactive labeling studies of the biosynthetic and transfer pathways of the cardiac glycosides in both the plants and the insects.

Feeding Experiments

Palatability of Monarchs reared on nonpoisonous plants

Brower and Brower (1964) proposed that one way of showing whether butterfly unpalatability is determined by their foodplants would be to select for a butterfly strain which could feed on a plant species belonging to a nonpoisonous group. During the fall of 1965 and the spring of 1966, we managed, with great difficulty, to produce a laboratory strain of Monarch butterflies which would eat cabbage (*Brassica oleracea* L.) (Brower, Brower, and Corvino, 1967). This cruciferous plant was chosen because it is known not to contain cardiac glycosides and can be easily grown in the greenhouse. Most butterflies reared on the cabbage leaves were small, weak and sterile, and very high mortality occurred at all stages, particularly from the 5th instar to pupa (see reference for details on how the stock was maintained). Our prediction was that Monarchs reared on this plant would be palatable to birds. In previous feeding experiments, scrub jays (*Aphelocoma coerulescens coerulescens* (Bosc), family Corvidae) occasionally attacked (and presumably tasted) but never ate Monarchs presented to them (J. Brower, 1958). In this experiment, wild-caught blue jays initially refused all Monarchs until placed upon an extreme food deprivation schedule. However, once the birds attacked the cabbage-reared Monarchs, they thereafter ate them freely without any signs of unpalatability (six birds were offered 38 insects and ate all but one).

Having thus broken down the reluctance of the birds to accept Monarchs, we next offered them material which had been reared on *Asclepias curassavica.* The birds ate from three-fourths to three individuals each, and in great contrast to their behavior after eating the cabbage-reared Monarchs, proceeded within about 12 minutes to enter a period of severe vomiting. Recovery followed within 20 minutes to an hour.

It was concluded from this experiment that the unpalatability of the Monarch butterfly is causally related to the species of plant ingested by the larvae. However, the possibility still exists that the insects synthesized the unpalatable components. It is, for example, possible that the cabbage-reared Monarchs were palatable because cabbage has within it chemical substances to which the Monarch is not adapted and which disrupt its metabolism, thereby preventing the insect from synthesizing its noxious properties. As we have already seen, the cabbage-reared butterflies were definitely abnormal.

Meanwhile, we had cultivated a second exotic milkweed from Trinidad in our Amherst College greenhouse, *Gonolobus rostratus* (Vahl) Roem. and Schult., and had reared Monarchs on this plant, too. Contrary to our expectation, these insects proved to be completely palatable to the birds. Professor Reichstein kindly analyzed dried material of this plant and reported to us that it is lacking in cardiac glycosides. Our field evidence from Trinidad indicated that immature Monarchs are occasionally found on *Gonolobus;* it is therefore a natural Asclepiad foodplant. Consequently, our experiment narrowed down the association of unpalatability in the Monarch to those Asclepiad plants containing cardiac glycosides as opposed to those lacking them and renders the alternative of synthesis less likely.

The palatability spectrum

The feeding experiments described above are qualitative in that they have shown whether Monarchs reared on various plants do or do not cause emesis and are acceptable or unacceptable to the birds. Parsons' most important assay experiments are limited in another way since they involved testing extracts of butterflies. It occurred to us that a more ecologically relevant bioassay would be one directed at determining that percentage of a whole butterfly necessary to cause emesis. Our aim, then, was to compare the emetic dosages of Monarchs reared on several species of Asclepiad plants.

Since these experiments have been described in more detail elsewhere (Brower, Ryerson, Coppinger, and Glazier, 1968), only a brief description of the technique and results will be given here. Monarchs were reared on *Asclepias curassavica* and two Old World species of Asclepiads known to contain considerable quantities of cardiac glycosides, *Calotropis procera* and *Gomphocarpus physocarpus* E. Meyer (or the closely allied *G. fruticosus* R. Br.). *Gonolobus rostratus*-reared Monarchs were used as controls. Adult males were killed by freezing, dried, and ground to a fine powder which was then loaded into gelatin capsules and force fed to blue jays. Dixon and Massey's (1957, Chapter 19) sequential experimental design was used to determine the emetic dose 50's of the three groups of insects.

The results are summarized in Table 1 where it can be seen that the E.D. 50 of the most poisonous Monarchs (those raised on *Calotropis*) is six times as great as the least (*Gomphocarpus*). If a blue jay were to eat one fifth of a *Calotropis*-reared butterfly, it would have a 50 percent chance of vomiting, whereas the bird would have to eat about 1.2 *Gomphocarpus* butterflies to be in the E.D.50 range. None of the control *Gonolobus*-fed birds vomited, so that the method *per se* did not evoke emesis.

These results confirm the previous findings in demonstrating that foodplant and palatability are causally related. But equally important, they also suggest that a spectrum of palatability occurs in nature in which some individuals are completely palatable, others barely tolerable, and still others too poisonous to be eaten even under situations of dire food shortage.

Table 1. Summary of Emetic Dose 50 Determinations by 80 Forced Feedings to 51 Blue Jays of Powdered Male Monarch Butterflies Reared as Larvae on Four Different Foodplants (From Brower et al., 1968)

Foodplant of butterflies	No. of tests	Dosage (in grams dry weight) of butterfly per 100 grams of bird		Mean No. E.D.50 blue jay units per butterfly[1]
		E.D. 50	95% Confidence limits	
Calotropis procera	20	.028	.024 to .033	4.8
Asclepias curassavica	20	.036	.032 to .040	3.7
Gomphocarpus sp.	20	.167	.148 to .188	.8
Gonolobus rostratus[2]	20	......	....	Not emetic

[1] Based on the mean dry weight of a sample of male Monarchs collected in Mayaro, Trinidad, W.I., in December 1967 (N = 50, Mean = .115g), and the average blue jay weight (N = 80, Mean = 85.43g).

[2] Dosages matched the *Gomphocarpus* dosages as a control against a large amount of material *per se* causing emesis. None of these birds vomited.

Implications for mimicry theory

When we discovered that *Gonolobus*-reared Monarchs were completely palatable, we proposed an additional theoretical category of mimicry: *automimicry* (Brower, Brower, and Corvino, 1967). In this phenomenon, a species is polymorphic for palatability which results from the larval foodplant selected by the ovipositing female. Since automimicry is intraspecific, the palatable individuals are perfect visual mimics of the unpalatable members of their own species. Predators will therefore be unable to discriminate between the two visually, and experience with the unpalatable individuals will result in subsequent avoidance of both. The advantage of the automimic will be frequency-dependent just as it is in Batesian mimicry. In other words, the palatable individuals cannot become too common relative to the unpalatable ones without losing the mimetic advantage (Brower, Cook, and Croze, 1967). The remarkable advantages of automimicry have recently been explored in more detail by Brower, Pough, and Meck (in press).

New Data

Support for automimicry has resulted from our earlier studies of breeding populations of Danaine butterflies in south-central Florida. In Highlands County two common species of milkweeds are *Asclepias humistrata* Walt. and *A. tuberosa rolfsii* (Britt.) Woodson. According to the review of Abisch and Reichstein (1962) *A. humistrata* contains cardiac glycosides, whereas *A. tuberosa* does not, although Punyarajun (1965) reports small amounts from the tuberous root. Brower (1962) collected a total of 270 eggs and larvae of the Monarch and another Danaine (*Danaus gilippus berenice* (Cramer), the Queen butterfly) on these two plants from one area over a two-week period. The results (Table 2) indicate that both species of insects are found upon

Table 2. Distribution of *Danaus plexippus* (Monarch) and *D. gilippus berenice* (Queen) Eggs and Larvae on *Asclepias humistrata* and *A. tuberosa rolfsii* on June 14, 21, and 28, 1960 in Area No. 1, Highlands County, Florida (From Brower, 1962)

	Monarchs	Queens	Totals
On *A. humistrata*	113 (.93)	88 (.59)	201 (.74)
On *A. tuberosa*	8 (.07)	61 (.41)	69 (.26)
Totals	121 (.45)	149 (.55)	270

Chi-square = 41.6; d.f. = 1; $P < .001$

the two plants, but to significantly different degrees. If the assumptions are made that *humistrata* produces unpalatable butterflies[1] and *tuberosa* palatable ones, and the Monarch and Queen are both able to assimilate the poisons, then both butterfly species populations are heterogeneous but the Queen more so than the Monarch.

To pursue this possibility, we have carried out voluntary feeding tests (as opposed to the forced-feeding technique described above) with Monarch butterflies reared in Florida on *Asclepias tuberosa rolfsii.* Storage of these butterflies was by our standard method: they are killed by freezing when less than 48 hours old and kept frozen in a deep freeze until used. Each butterfly is in an individual glassine envelope and each lot of butterflies is placed in a plastic bag from which the air is squeezed out before the top is secured. Immediately before use, the butterflies are removed and thawed out.

Ten wild-captured adult blue jays from Hampshire and Franklin counties, Massachusetts, were individually caged and induced first to eat a standard palatable butterfly (*Anartia amalthea* L., see Brower and Brower, 1964) by means of food deprivation and our new rotary feeding devices (see Brower, Brower, and Corvino, 1967). Once the birds began to eat *Anartia* freely, they were then offered palatable *Gonolobus*-reared Monarch butterflies sequentially and singly for two days, followed by *Asclepias tuberosa*-reared butterflies on two more days (Table 3). The results are clear-cut. The ten birds ate 37 out of 40 of the *tuberosa* butterflies offered on the first day and all 20 offered one day later. Neither emesis nor signs of sickness were observed, and the acceptance of the butterflies on the final day clearly indicates that the birds found them palatable. This experiment is consistent with the reported absence of heart poisons in the leaves of *A. tuberosa.* (But see Duffey, 1970, Science 169:78.)

Although more feeding experiments need to be done, it is now virtually certain that the Monarch butterfly exhibits a palatability polymorphism and automimicry in nature which is determined by the distribution of poisonous and nonpoisonous foodplants and female oviposition choice (see also Brower, 1969). The geographic distribution of *A. tuberosa* is well known (Woodson, 1954) and is shown in Figure 1. The distribution of the poisonous *A. humistrata* is shown in Figure 2.

Combining the fact of the vast geographic distribution of the Monarch butterfly in North America mediated by its migratory be-

[1] *Note added in proof:* Experiments conducted since this paper was written indicate that Monarchs reared on *A. humistrata* are more poisonous than those reared on *A. curassavica* (see Brower, 1969).

Table 3. Reactions of Ten Individually Caged Blue Jays[1] to Monarch Butterflies[2] (*Danaus plexippus*) Reared on Two Asclepiad Plants, *Gonolobus rostratus* and *Asclepias tuberosa rolfsii*; No Vomiting and No Indications of Unpalatability Were Observed[3]

		No. butterflies[4] offered and eaten by bird no.:										
Day[5]	Larval foodplant	1	2	3	4	5	6	7	8	9	10	Totals
1	*Gonolobus*	2/2	2/2	2/2	2/2	2/2	2/2	2/2	2/2	2/2	2/2	20/20
2	*Gonolobus*	2/2	2/2	2/2	2/2	2/2	2/2	2/2	2/2	2/2	2/2	20/20
3	*A. tuberosa*	4/4	3/4	4/4[6]	3/4	4/4	4/4	4/4	3/4	4/4	4/4	37/40
4	*A. tuberosa*	2/2	2/2	2/2	2/2	2/2	2/2	2/2	2/2	2/2	2/2	20/20

[1] Birds wild-captured in fall 1967, Franklin and Hampshire counties, Mass.
[2] Butterflies reared from separate stocks of southern Florida origin. *Gonolobus* cultivated in Amherst College greenhouse of Trinidad, W.I., origin. *Asclepias tuberosa* collected in wild in southern Florida (Highlands County.)
[3] All birds were observed for at least 30 minutes following ingestion of the first butterflies on days 2, 3, and 4.
[4] *Gonolobus*-reared females offered on day 1, males on day 2. *Asclepias*-reared males offered to birds 1-5, females to birds 9 and 10. Birds 6-8 received both sexes.
[5] Experiment conducted 10 November to 15 December 1967. Days 1-4 all consecutive except for birds 1 and 3 in which 1-2 days intervened between days 1-2 and 2-3.
[6] Partial regurgitation of material occurred after eating third butterfly.

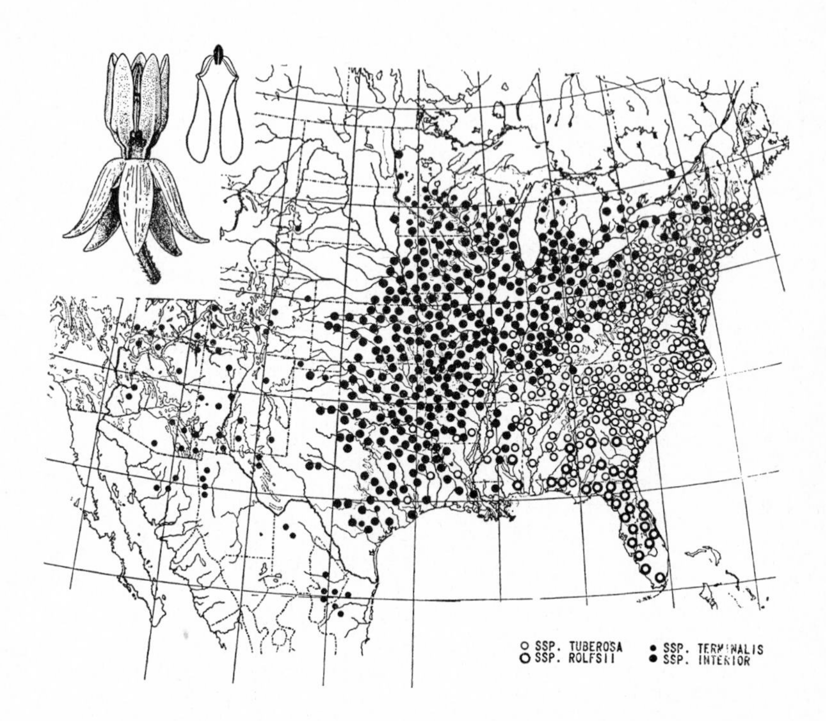

FIGURE 1. Geographic distribution of *Asclepias tuberosa*. This milkweed is a widespread larval foodplant of the Monarch butterfly. Monarchs which have eaten this plant are highly palatable to birds (Map from Woodson (1954); reproduced by permission of the Missouri Botanical Garden.)

havior (Urquhart, 1960) with the fact that there are 106 species of known North American *Asclepias* in addition to *A. tuberosa* and *A. humistrata* (Woodson, 1954), one can readily appreciate that it is most likely that the palatability heterogeneity of Monarch populations is constantly changing in both space and time. It is therefore not surprising that some investigators (e.g., Petersen, 1964) have reported that birds do eat Monarch butterflies, and others the opposite. Undoubtedly, both groups of investigators are correct: some Monarchs are palatable, some are not, and others are intermediate, and the frequencies probably shift from one extreme to the other in different areas, different seasons, different years, and perhaps even from day to day during migratory periods.

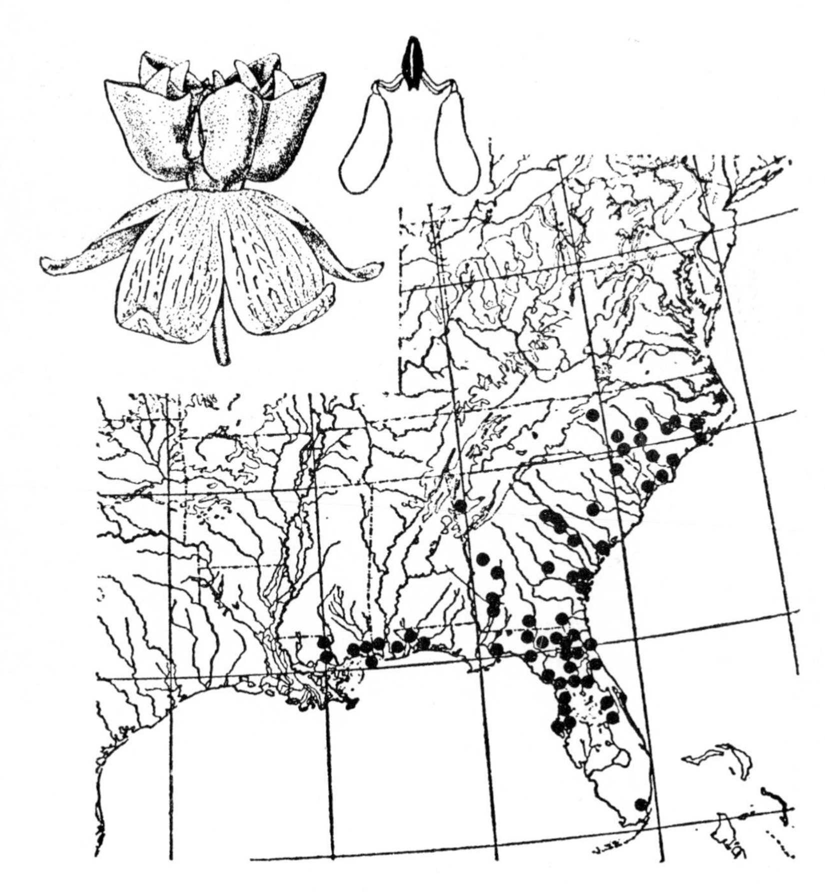

FIGURE 2. Geographic distribution of *Asclepias humistrata*. This milkweed contains cardiac glycosides and is a larval foodplant of the Monarch butterfly in south-central Florida. Monarchs which have eaten this plant are very unpalatable to birds and cause severe vomiting. (Map from Woodson (1954); reproduced by permission of the Missouri Botanical Garden.)

Conclusions

The likelihood that Danaine butterflies are able to assimilate cardiac poisons from their foodplants is now firmly established and is a remarkable example of biochemical coevolution. Since the cardiac glycoside content of various species of Asclepiad plants varies both qualitatively and quantitatively (from none to large amounts), the palatability of these insects to vertebrate predators must consist of a spectrum from completely acceptable to completely unacceptable.

As a consequence, the new concept of automimicry in Danaine butterflies is now a virtual certainty and probably also occurs in other insect groups. Not only do automimicry and the palatability spectrum provide a basis for understanding the controversies that have existed for so long about warning coloration, palatability, and mimicry, but they also will prove powerful tools for further delving into the complex interactions within plant-herbivore-predator systems.

ACKNOWLEDGMENTS: I am indebted to many who have helped with various aspects of this work and particularly Professor Tadeus Reichstein, professor of Organic Chemistry at Basel University. I also wish to thank Miriam Rothschild for suggesting our valuable collaboration. William Ryerson, an Amherst College senior honor's student now studying for his Ph.D. at Yale University, carried out a brilliant investigation which served as the basis for the new ecologically oriented emetic dosage 50 technique. I am also grateful to Professor James Q. Denton of Amherst College for mathematical advice, and to Professor E. B. Ford, F.R.S., and Dr. Jane VanZandt Brower for criticizing the manuscript. Miss Judith Myers made it possible to do the feeding experiments by raising the *Asclepias tuberosa* Monarch butterflies in Florida. The competent assistance of Raymond P. Coppinger, Lorna L. Coppinger, and Susan C. Glazier in the experimental work deserves hearty thanks. I would also like to thank Mrs. Helen Sullivan for her patience and most careful typing. Portions of the investigation were carried out at the William Beebe Tropical Research Station in Trinidad, W.I., and the Archbold Biological Station in Florida. The work was made possible by National Science Foundation grants GB-4924 and GB-7637.

Literature Cited

Abisch, Eva, and T. Reichstein. 1962. Orientierende chemische Untersuchung einiger Asclepiadaceen und Periplocaceen. Helv. Chim. Acta *45:* 2090-2116.

Brower, Jane V.Z. 1958. Experimental studies of mimicry in some North American butterflies. Part 1. The Monarch, *Danaus plexippus,* and Viceroy, *Limenitis archippus archippus.* Evolution *12:* 32-47.

Brower, L. P. 1961. Studies on the migration of the Monarch butterfly. 1. Breeding populations of *Danaus plexippus* and *D. gilippus berenice* in south-central Florida. Ecology *42:* 76-83.

Brower, L. P. 1962. Evidence for interspecific competition in natural populations of the Monarch and Queen butterflies, *Danaus plexippus* and *D. gilippus berenice* in south-central Florida. Ecology *43:* 549-552.

Brower, L. P., and J.V.Z. Brower. 1964. Birds, butterflies, and plant poisons: a study in ecological chemistry. Zoologica *49:* 137-159.

Brower, L. P., J.V.Z. Brower, and J. M. Corvino. 1967. Plant poisons in a terrestrial food chain. Proc. Nat. Acad. Sci. *57:* 893-898.

Brower, L. P., L. M. Cook, and H. J. Croze. 1967. Predator responses to artificial Batesian mimics released in a neotropical environment. Evolution *21:* 13-21.

Brower, L. P., W. N. Ryerson, Lorna L. Coppinger, and Susan C. Glazier. 1968. Ecological chemistry and the palatability spectrum. Science *161:* 1349-1351.

Brower, L. P. 1969. Ecological chemistry. Sci. Am. *220* (*2*): 22-29.

Brower, L. P., F. H. Pough, and H. R. Meck. In press. Theoretical investigations of automimicry. 1. Single trial learning. Proc. Nat. Acad. Sci. *65.*

Dixon, W. J., and F. J. Massey, Jr. 1957. *Introduction to Statistical Analysis,* 2nd edition. New York: McGraw Hill Book Co.

Duggan, R. E., and J. R. Weatherwax. 1967. Dietary intake of pesticide chemicals. Science *157:* 1006-1010.

Ehrlich, P. R., and P. H. Raven. 1965. Butterflies and plants: a study in coevolution. Evolution *18:* 586-608.

Ehrlich, P. R., and P. H. Raven. 1967. Butterflies and plants. Sci. Am. *216* (6): 104-113.

Eisner, T., and J. Meinwald. 1966. Defensive secretions of arthropods. Science *153:* 1341-1350.

Euw, J. V., L. Fishelson, J. A. Parsons, T. Reichstein, and Miriam Rothschild. 1967. Cardenolides (heart poisons) in a grasshopper feeding on milkweeds. Nature *214:* 35-39.

Gero, A. 1965. Cardiac glycosides 1: chemistry. In *Drill's Pharmacology in Medicine,* 3rd ed., J. R. DiPalma, ed. New York: McGraw Hill, pp. 567-572.

Hoch, J. H. 1961. *A Survey of the Cardiac Glycosides and Genins.* Columbia: University of South Carolina Press.

Moe, G. K., and A. H. Farah. 1965. Digitalis and allied cardiac glycosides. In *The Pharmacological Basis of Therapeutics,* 3rd ed., L. S. Goodman and A. Gilman, eds. New York: Macmillan, pp. 665-698.

Parsons, J. A. 1965. A digitalis-like toxin in the Monarch butterfly, *Danaus plexippus* L. J. Physiol. *178:* 290-304.

Petersen, B. 1964. Monarch butterflies are eaten by birds. J. Lepidopterists' Soc. *18:* 165-169.

Punyarajun, Sasri. 1965. Chemical investigation of *Asclepias tuberosa.* University Microfilms, Inc., Ann Arbor, Michigan. Ph.D. thesis.

Reichstein, T. 1967. Cardenolide (herzwirksame Glykoside) als Abwehrstoffe bei Insekten. Naturwissenschaftliche Rundschau *20:* 499-511.

Slater, J. W. 1877. On the food of gaily-coloured caterpillars. Trans. Ent. Soc., London *1877:* 205-209.

Urquhart, F. A. 1960. *The Monarch Butterfly.* Toronto: University of Toronto Press.

Woodson, R. E., Jr. 1954. The North American species of *Asclepias* L. Ann. Missouri Bot. Gard. *41* (*1*): 1-211.

The Role of Chemical Attractants in Orchid Pollination

C. H. DODSON
Department of Biology
University of Miami

EVEN THE MOST CONSERVATIVE estimates place the orchid family far ahead of other families of flowering plants in numbers of species. Airy-Shaw, in the new edition of Willis (1967), estimates 20,000 species for the Orchidaceae versus 12,000 for the Compositae, the next largest family. We estimate (van der Pijl and Dodson, 1966) that 10 percent of the species of flowering plants are orchids.

Why should one family have speciated so explosively when most families of flowering plants have considerably less than 500 species? I believe that the orchids, through the development of morphological structures in the flower and complex reproductive systems coupled with adaptation to diverse habitats, have not only lent themselves to rapid speciation but have failed to elicit the stringent extinction which has occurred in other groups (Dressler and Dodson, 1960). Consequently, new taxa develop while old taxa remain on the scene. Sympatric speciation becomes commonplace rather than of rare occurrence (Dodson, 1962; Dressler, 1968b). Genetic differences are not developed with sufficient rapidity to preclude hybridization (under experimental conditions) as has been overwhelmingly demonstrated by the orchid-growing industry. Even genera can be crossed with relative ease and they are clearly valid genera, not the "fictitious genera of the orchid taxonomist" which have become notorious in taxonomic circles.

In some cases, interfertile orchid genera are sufficiently diverse in their morphological characters that they would easily qualify as distinct subfamilies or even families were they in the Scrophulariales or Sapindales. Curiously enough, records of natural hybrids between genera in the orchids are exceedingly rare, and those which have been reported are usually between genera of a genuinely controversial nature.

If the above is true, there must be a valid reason why numerous species have evolved and why they do not tend to be swamped by the effects of widespread hybridization in their natural habitat. The fact that they show an inordinate interfertility in the greenhouse should not cast doubt on their validity as taxa if they tend not to hybridize in nature. With such large numbers of species—some genera have more than a thousand valid species—it would seem reasonable that many instances of sympatric, interfertile species would occur. In truth, they do occur together. Six species of *Stanhopea* occur on the road cuts near Turrialba, Costa Rica. Five species of *Coryanthes* occur in trees along the Amazon River at Iquitos, Peru. All of the species of *Stanhopea* are interfertile as are the *Coryanthes* (Dodson, 1965b). Two of the species of *Stanhopea* at Turrialba form rare natural hybrids which do not backcross because of mechanical irregularities. Two other species at the same locality hybridize occasionally (Dodson, 1965a). No hybrids were found between the *Coryanthes* at Iquitos. Numerous similar instances could be cited.

The reason for the high number of orchid species seems to me to be primarily due to the development of reproductive isolating mechanisms based on remarkably specific attraction of pollinators. Orchids have adapted to most of the available pollinating agents in the environment, with the notable exception of wind, water, and bats. None of these agents are amenable to operating within the structural limits of the orchid flower. We have estimated the percentage of orchid species pollinated by the various agents (van der Pijl and Dodson, 1966), based on field observations of morphological adaptations of the orchids to particular kinds of pollinators. We are thereby able to predict the flower class to which a given orchid species might belong with a relative degree of accuracy. Table 1 is included to illustrate the pollinating agents and the percentages of orchid species adapted to them.

In this paper, only one flower class will be discussed. An estimated 2,000 orchid species, restricted to the tropical regions of the Western Hemisphere, are adapted exclusively to pollination by the golden bees (Euglossini). The flowers of some of these orchids are among the most bizarre to be found among flowering plants, and the complex structures of the flowers lend themselves to mechanical isolation of species. The orchids exclusively adapted to euglossine pollination tend to form natural groups, for example, all members of the subtribes Catasetinae, Lycastinae, Stanhopinae, and Zygopetalinae and some genera on the Oncidinae, such as *Aspasia, Notylia, Rodriguezia, Trichocentrum,* and *Trichopilia* (Dodson, 1965a; van der Pijl and Dodson, 1966; Dressler; 1968b). Most of the species in these genera attract bees quite specifically and in many cases only one species of bee

Table 1. A Pollination Spectrum for the Orchid Family Showing the Percentage of Orchid Species Adapted to Particular Pollination Agents (From van der Pijl and Dodson, 1966)

Hymenoptera		Other agents	
	%		%
Wasps	5	Moths	8
Lower bees	16	Butterflies	3
Carpenter bees	11	Birds	3
Euglossini	10	Flies	15
Social bees	8	Mixed agents	8
Mixed bees	10	Autogamous	3
	—		—
Totals	60		40

visits a particular orchid species (Dodson, 1962). Curiously, only golden bee males are attracted to these orchid flowers, and they are attracted primarily to the fragrances produced (Adams, 1966). The euglossine-flower (flowers pollinated by male euglossine bees) in the orchids does not produce nectar or other food for the bees.

The Golden Bees

The Euglossini is a group of long-tongued, tropical bees with more than 200 species. Their similarities are many and they are considered to form a separate tribe of the Apidae (Moure, 1950) composed of six genera: *Euglossa, Eulaema, Euplusia, Eufriesia, Exaerete,* and *Aglae. Eufriesia, Exaerete,* and *Aglae* appear to be rarely involved with orchid flowers and comprise only seven or eight species. *Eulaema* consists of about 17 species of large, hairy bees with black heads and thoraxes. The color of the abdomen varies from species to species; some are black with a yellow or brown tip, while most are orange-yellow or banded with yellow, black, or red. *Euplusia* consists of about 40 species of medium to large bees which closely resemble the species of *Eulaema* in most cases; however, all have a brilliantly metalic green or blue head and thorax among other characters which separate the genera. *Euglossa* consists of more than a hundred species (the number of recognized species in this genus has nearly doubled since we began our study in 1959) of small to medium size, brilliantly metallic blue, green, or golden bees (it is from this genus that the common name "golden bees" was derived) which are sparsely hairy. All of the bees are rapid fliers and tend to be very wary. They do not occur outside the tropical zones of the Western Hemisphere. Some species are very local in distribution, while others occur from central Mexico to southern Brazil.

The behavior of *Euglossa, Eulaema, Euplusia,* and *Eufriesia* is for the most part similar (Dodson, 1966). *Exaerete* and *Aglae* are parasitic bees which lay their eggs in nests of other members of the tribe. The female bees of *Euglossa, Eulaema, Euplusia,* and *Eufriesia* forage in the field from dawn to dusk, collecting construction materials and food for storage in the cells of their nests. The females occasionally visit orchid flowers which are pollinated by other bees, such as *Bombus* and *Xylocopa,* but they never visit the euglossine-flowers which attract males exclusively. There appears to be little division of labor, each female constructing her own nest and cells in most cases. In some species of *Eulaema* and *Euglossa,* several females may occupy a single large nest and apparently cooperate in its construction and provisioning (Dodson, 1966; Roberts and Dodson, 1967). The female bees are to be seen at all hours of the day visiting local plants for pollen and nectar. They tend to frequent tubular flowers with long nectaries, such as members of the Marantaceae, Zingiberaceae, Apocynaceae, Bignoniaceae, and Gesneriaceae from which nectar can be collected with their long tongues.

The male bees leave the nest immediately upon emerging and do not return to the nest. They seem to live a vagabond life, feeding on various nectar-producing flowers and visiting orchid flowers (and flowers of some members of the Araceae, Gesneriaceae, Solanaceae, and Lecythidaceae). The male bees are long lived—records of marked male bees living six months have been reported (Dodson, 1962). The male bee establishes a territory which it defends and patrols in a manner typical of its species. Some choose a limb or a tree trunk and land on it for a moment and then fly a prescribed pattern before returning and landing again. This behavior may continue for hours. Apparently, an odor is left at the spot where landing takes place, for if one male bee is captured a new male frequently takes its place at the same spot the following day. The female bee, ready for mating, is attracted either by the flight or by the odor deposited by the male. Copulation takes place at the site (Dodson, 1966).

Male euglossine bees differ morphologically from all other bees by the presence of brushes on the tarsi of the front legs and the swollen tibeae of the rear legs (Figure 1). The swollen tibea contains glands which are associated with an elongate opening termed the "scar" on the rear of the tibea (Cruz, et al. 1965; Sakagami, 1965). The "scar" is surrounded by stiff hairs.

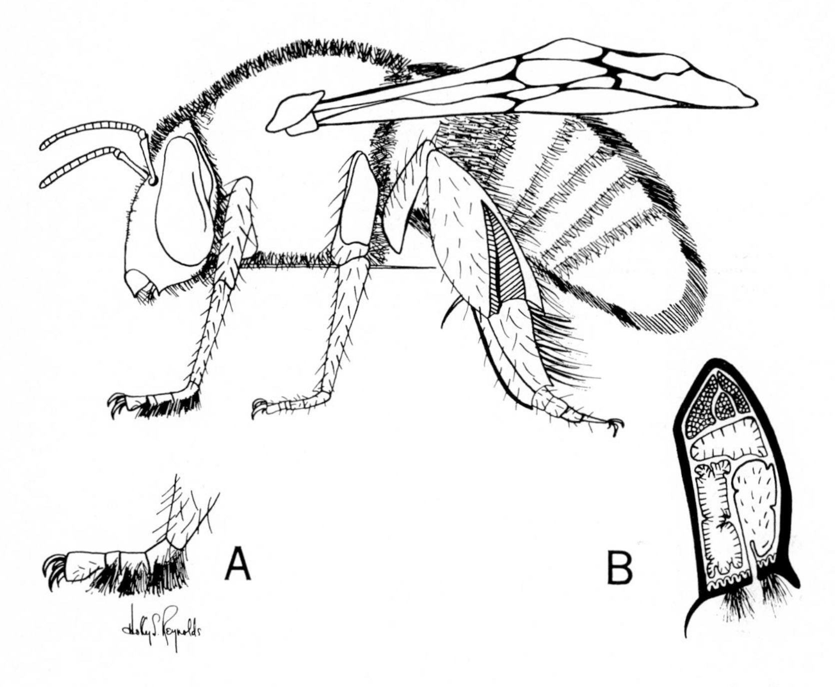

FIGURE 1. Male euglossine bee. A. Tarsus of the front leg showing brushes; B. section of tibia of hind leg showing structure.

The Visits of Male Bees to Orchid Flowers

Golden bees have been reported as pollinators of orchids since 1865 (Cruger, 1865; Ducke, 1902; Allen, 1951, 1952, 1954; Porsch, 1955). In all of these reports it was assumed that the bees came to the flowers for food, and since no nectar was evident it was postulated that the bees gnawed the floral tissues. Dodson and Frymire (1961) and Vogel (1963) reported detailed observations of pollination of euglossine orchids in which it was clear that the bees were strictly males and that they did not consume tissues. While visiting the flower, they behave in a characteristic manner. The bees are very wary on arrival at the flower, hovering and darting away only to return and hover again while testing the fragrance. When they land on the flower, they immediately commence rubbing the surface of the lip with their tarsal brushes. After 30 to 60 seconds, the bee again resumes flight and hovers in front of the flower, rubbing the tarsal pads against the scars on the swollen tibea. It then returns to the flower and repeats the process.

After the bee makes the first landing, it ceases to be wary and can be approached rather easily. Earlier, the author believed that the bee became "intoxicated" by the materials rubbed from the flowers and transferred to the tibeae of the legs (Dodson and Frymire, 1961; Dodson, 1962). My recent observations of female euglossine bees collecting resin for nest construction (from a tree of *Proteum*) indicate that the female bees react in the same manner after beginning the removal of the resin and placing it in the pollen baskets. The resin probably does not "intoxicate" the female bee. Therefore, it seems more likely that the "drugged" aspect of the behavior of both male and female euglossine bees results from "concentration" rather than a physiological reaction to chemical compounds produced by the flowers. Vogel (1966) has demonstrated through the use of thin-layer chromatography that the materials stored in the hind tibeae are the same as the materials on the surface of the flower. We have demonstrated through the use of gas-liquid chromatography that many of the odorous compounds contained in the tibeae after a visit to an orchid flower are the same as those in the fragrance of the flower.

While in this state—whether it results from "intoxication" or from "concentration and a one-track mind"—the male bee falls easily, does not fly well, and is generally impaired in motor response. Some orchids have adapted by producing pollination mechanisms which depend on this characteristic behavior. When a bee launchs itself into the air in order to transfer the fragrance material from the tarsal pads to the tibeal glands, it falls. During this time, the bee can be "manipulated" easily by complex flowers.

Some euglossine-orchids are relatively simple (Figure 2) and are visited and pollinated much as in other orchid flowers. The pollinaria (pollen masses and associated structures which attach them to the pollinator) are attached to the bee in quite specific places, to be transported to the stigma of another flower of the same species. By varying the attachment point of the pollinarium (e.g., behind the thorax, on top of the thorax, behind the head, on the front of the head, under the abdomen, on the antennae, on the legs, on the tongue, and so forth) and concurrent positioning of the stigma in relation to the posture of the visiting bee, it is possible for several sympatric orchid species to be pollinated by the same species of bee without danger of hybridization (Dodson, 1965; Dressler, 1968a). More complex flowers (e.g., *Gongora* spp. and *Stanhopea* spp.) depend upon a bee falling through the flower after landing on the exposed base of the lip (Figure 3). A different mechanism in the genus *Coryanthes* involves the bee falling into a bucket partially filled with water (Figure 4). Here the apical portion of the lip of the flower forms the bucket while water drips

FIGURE 2. Male *Eulaema cingulata* visiting *Trichocentrum tigrinum.* An example of a relatively simple orchid flower without complex pollination mechanisms.

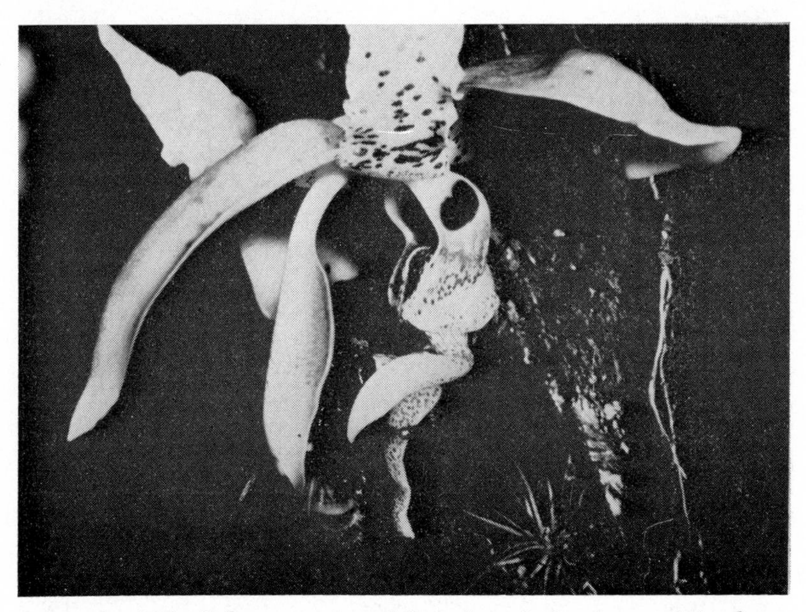

FIGURE 3. Male *Eulaema meriana* falling through the flower of *Stanhopea gibbosa.*

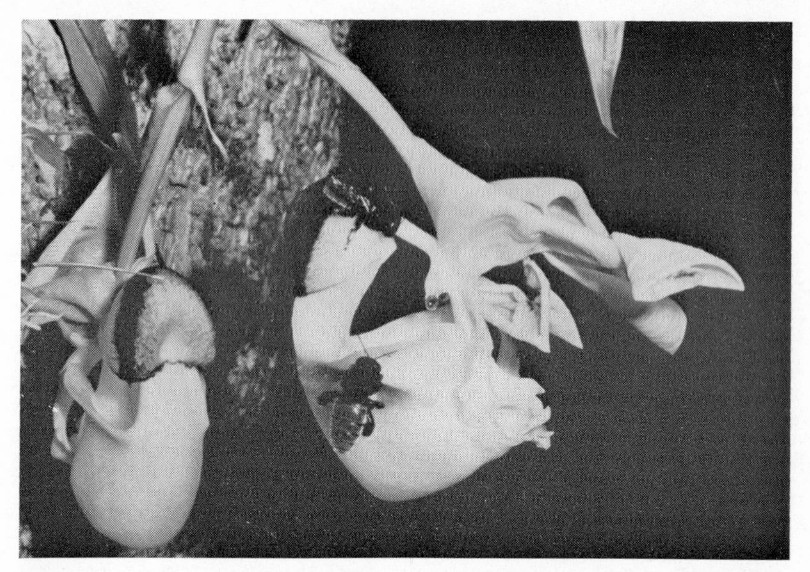

A

B

FIGURE 4. A. Male *Euplusia superba* visiting *Coryanthes rodriguezii;* B. male *Euplusia superba* crawling from the opening under the anther and stigma after falling into the water-filled, bucket-like apical portion of the lip of the flower.

from glands on each side of the base of the flower. The bee falls into the water in the bucket and in order to escape from the flower must crawl under the stigma and anther. Such complex mechanisms would not function with "alert" bees.

In many cases dimensions of the flower act as isolating mechanisms through adaptation to small or large bees, excluding bees not of proper dimensions. Morphological isolating mechanisms, however, are not sufficient to reduce the numbers of different species of the same size. At any given site, numerous species of euglossine bees may be present. For example, 57 species are known for central Panama and probably 40 of them could be found at any specific locality in that area. Therefore, even a combination of adaptation to differential size of pollinator and placement of pollinia would be insufficient to provide the specificity necessary to keep numerous interfertile species from hybridizing and thereby loosing their integrity. The factor of differential production of chemical components in the fragrances produced, coupled with strong attraction of euglossine bees to specific fragrances, provides the necessary specificity to allow for the existing situation.

Specific Attraction to Chemical Components of Floral Fragrances

It becomes obvious, even to the casual observer in the field, that male euglossine bees are strongly attracted to the floral fragrances produced by euglossine-orchids. Even if the flowers are hidden from view, the bees seek them out. Adams (1966) placed flowers of *Catasetum maculatum* inside a clear plastic insect trap, left the flowers for a few minutes, then removed the flowers. Bees were attracted to the residual odor in the trap. Records of intensive field observations of euglossine pollination of orchid flowers (Dodson and Frymire, 1961; Dodson, 1962, 1965a; Dressler, 1968b) over several years clearly demonstrate that the significant means of selective attraction of euglossine bees resides in differential production of fragrance components.

Laboratory identification of attractants

Gas-liquid chromatography was employed to analyze the floral fragrances qualitatively (Dodson and Hills, 1966). The results of the analysis of floral fragrances of approximately 150 species of orchids from 16 genera, which are known to be euglossine pollinated, indicate that a total of approximately 60 chemical compounds can be present (Table 2). Most species produce from 7 to 10 compounds, with some having as many as 18 and others as few as 2 (Figure 5). The relative retention times of the compounds were calculated from numerous runs

Table 2. A Comparison of Some of the Compounds Present in Orchid Flower Fragrances

Compounds \ Orchid species	*Brassavola digbyana*	*Catasetum fimbriatum*	*C. macrocarpum*	*C. ochraceum*	*C. roseum*	*C. tenebrosum*	*C. thylaciochilum*	*C. warscewiczii*	*Coryanthes macrantha*	*Stanhopea anfracta*	*S. annulata*	*S. florida*	*S. grandiflora*	*S. pulla*	*S. tigrina*
0.43[1]			X	X											
0.63 alpha-pinene		X	X		X	X		X		X	X	X	X	X	
0.68	X	X													
0.78			X												
1.00 beta-pinene		X	X		X				X			X		X	
1.06			X		X	X		X		X	X				
1.33 Myrcene		X		X	X			X	X	X	X	X	X		
1.44 alpha-phellandrene														X	
1.68 Limonene			X		X	X								X	
1.80 1,8-cineole			X		X			X		X	X	X	X	X	
2.14											X				
2.23 Ocimine	X			X		X				X					
2.38 p-cymene		X									X			X	
2.57			X			X				X		X			
0.42 Citronellal			X									X		X	X
0.53 Linalool	X	X						X		X	X	X		X	

0.58			X		X								X		
0.70 Geraniol						X		X		X	X				X
0.74 Caryophyllene						X									
0.84 Methyl benzoate			X		X										
0.96			X			X		X		X	X				
1.05 alpha-terpineol (?)					X							X			
1.27 Benzyl acetate			X							X		X	X	X	X
1.30 Piperitone									X						
1.31 d-carvone			X												
1.34															
1.36 Citronellol	X														
1.41				X											
1.56 Methyl salicylate										X		X			
1.71			X											X	
1.84 2-phenylethylacetate							X								X
2.00		X													
2.08 Nerol								X							
2.31													X		
2.67 2-phenylethanol							X								
2.75															X
3.54				X											
5.35 Methyl cinnamate							X								

[1] Relative retention times in minutes.

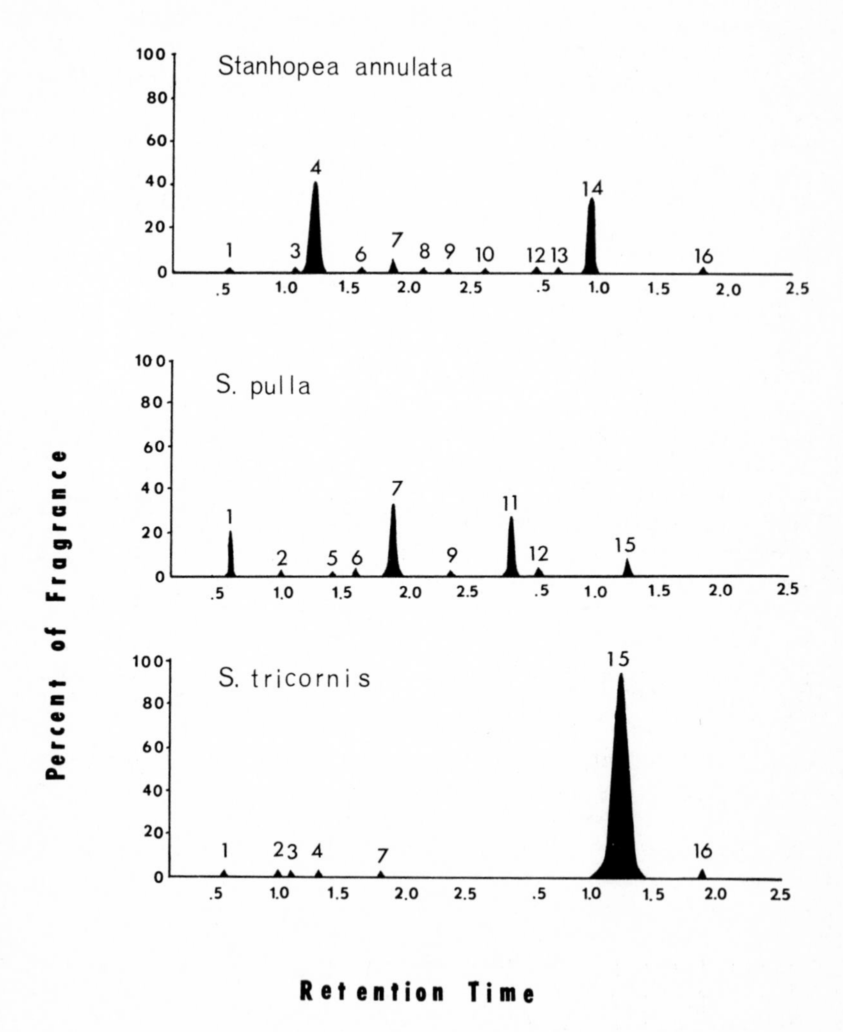

FIGURE 5. Comparison of chromatographs of three species of *Stanhopea.*

on two different columns (3% carbowax 20 M and 10% Lac 446) at different oven temperatures (70° and 130° C). For example, a peak which consistently appears at 1.80 minutes (relative retention time at 70° C on the carbowax column, under our techniques) is found in approximately 60 percent of the species sampled. Some compounds occur more frequently, but most of them occur less frequently (Table 2).

The compounds are produced in different proportions in different species; for example, the 1.80 compound forms about 90 percent of the total fragrance produced in *Stanhopea cirrhata,* while the same compound only forms about 7 percent of the total fragrance of *Catasetum maculatum.*

The chemical identities of the compounds were unknown until late 1967 when we began to direct our efforts toward the identification of certain of the more widely occurring compounds. *Catasetum collare* produces a fragrance similar to methyl salicylate (oil of wintergreen). Laboratory-produced methyl salicylate has a relative retention time of 1.57 minutes at 130° C on carbowax under our techniques. The fragrance of the flower of *C. collare* is dominated by a peak with a retention time of 1.57 minutes. The fragrance of *Stanhopea cirrhata* is reminiscent of Vicks Vaporub. One of the major constituents of Vicks Vaporub is cineole (Eucalyptol) which has a relative retention time of 1.80 minutes at 70° C on carbowax. The major component of the fragrance of *Stanhopea tricornis* has a relative retention time of 1.27 minutes at 130° C on carbowax and smells similar to benzyl acetate. Benzyl acetate has the same relative retention time.

The three compounds, cineole, methyl salicylate, and benzyl acetate were further tested in the laboratory by comparing their retention times against a standard compound (beta pinene at 70° C and ethyl benzoate at 130° C) on the two columns. In all cases the retention times agreed with those of the unknown compounds from the flower and smelled the same as those compounds as they leave the heated collection vent. The known compounds were then run as a mixture with the orchid fragrances at a reduced temperature, and they did not produce double peaks with the compounds in question. The unknown compounds were then considered to be tentatively identified and ready for field testing to determine their reaction with the bees that pollinate the orchids. Several other compounds present in the orchid fragrances were then identified by the same procedures. These include alpha-pinene, beta-pinene, linalool, methyl benzoate, d-carvone, 2-phenylethanol, and citronellol.

Field testing

Field tests of the effect of the chemical compounds were made in February of 1968. The area chosen for the tests was the central portion of Panama on both sides of the Canal Zone. Elevations and ecological zonation of the three sites, Cerro Jefe east of Panama City, Santa Rita Ridge east of Colon, and Cerro Campana west of Panama City, are relatively similar. Elevations range from 700 to 1,300 meters, and all three sites were known to have numerous euglossine bees present.

The tests were made by saturating 2 by 2-inch squares of blotter paper with 1 ml of the material to be tested (Figure 6). These squares were tacked to tree trunks or logs in natural forest areas. The blotters were spaced from two to five meters apart. The results of five days of trials are included in Tables 2, 3, 4, and 5. The materials were tested between 7 a.m. and 1 p.m., and the bees attracted were captured in insect nets and killed in order to determine species and to assure that they were not counted twice. When the materials appeared to have evaporated somewhat from the pads, they were replenished. Some compounds were quite volatile, others would last most of the day without replenishing.

Fifty-seven species of euglossine bees are known from this area. These include 36 species of *Euglossa,* 8 species of *Eulaema,* 10 species of *Euplusia,* 1 species of *Eufriesia,* and 2 species of *Exaerete.* Of these, 28 species of *Euglossa,* 6 species of *Eulaema,* 5 species of *Euplusia,* 1 *Eufriesia,* and 2 species of *Exaerete* were attracted (see Table 3).

Table 3. NUMBERS OF SPECIES FROM EACH EUGLOSSINE GENUS ATTRACTED TO THE VARIOUS COMPOUNDS TESTED

Compound	*Euglossa*	*Eulaema*	*Euplusia*	*Eufriesia*	*Exaerete*
Cineole	27	4	2		2
Methyl salicylate	7	2	1	1	
Benzyl acetate	2	3	1		
d-carvone			1		
2-phenylethanol	1				
2-phenylethylacetate			1		
Methyl benzoate		2			
Linalool		1			

A total of 42 of the 57 species known from central Panama were attracted. Some of the 15 unattracted species were obviously present at the time the tests were made and were simply not attracted by the compounds put out. Others are probably seasonal or extremely rare and may not have been present at the time of the tests.

Two kinds of trials were made. Pure compounds and mixtures of compounds were put out. Some of the compounds in pure form acted as general attractants (e.g., cineole). Some compounds acted as specific attractants (e.g., d-carvone and 2-phenylethanol) and some failed to attract. Various mixtures of cineole, benzyl acetate, alpha pinene, and beta pinene, approximately in the relative concentrations found in orchid flowers, were also tested and their attractiveness recorded (Table 5).

A

B

FIGURE 6. A. Male bees of *Eulaema nigrita, Euglossa asarophora,* and *Euglossa dodsonii* visiting a piece of blotter paper saturated with cineole; B. *Eulaema meriana* visiting a piece of blotter paper saturated with one part cineole and twenty parts benzyl acetate.

Table 4. RESULTS OF FIELD TESTS OF NINE CHEMICAL ATTRACTANTS OF MALE EUGLOSSINE BEES AT THREE SITES IN CENTRAL PANAMA (Cerro Jefe, Cerro Campana, and Santa Rita Ridge)[1]

	Cineole	Methyl salicylate	Benzyl acetate	Methyl benzoate	d-carvone	Linalool	2-phenyleth-anol	Alpha ionone	Beta ionone
Euglossa allosticta Moure ined.	18								
E. asarophora Moure	76								6
E. azureoviridis Friese	1								
E. bursigera Moure	9								
E. championi Cheesman	26	15							
E. cyanaspis Moure	2								
E. cyanura Cockerell							3		
E. cybelia Moure	9	4		1					
E. deceptrix Moure	34								
E. dodsoni Moure	24		6						
E. dressleri Moure	11		1						
E. flammea Moure	3								
E. gorgonensis Cheesman	17								
E. hansoni Moure	13								
E. ignita Smith	3								
E. igniventris Friese	30								
E. imperialis Cockerell	48	39							
E. maculilabris Moure	16								
E. mixta Friese	2	7							
E. sapphirina Moure	5	24							
E. variabilis Friese	16								
E. tridentata Moure	35	1							
E. sp. (RD 120)	1								
E. sp. (RD 355)	9								
Eulaema bombiformis (Packard)		11		3					
E. cingulata Fabricius			17					2	5
E. meriana Oliver		2	12						
E. nigrita Lepeletier	9					1			
E. polychroma Friese	2								2
E. speciosa Mocsary	4								1
Euplusia concava Friese	1				2				
E. sp. (RD 237)		11							4
Eufriesia pulchra Smith		9							
Total bees—612[2]	433	113	36	4	2	1	3	2	18

[1] Tests were made during five days late in January 1968.
[2] Several of bee names included above have not been published but represent valid species.

Table 5. RESULTS OF FIELD TESTS OF COMBINATIONS OF FOUR CHEMICAL COMPOUNDS WHICH AFFECT ATTRACTION OF MALE EUGLOSSINE BEES[1]

	Cineole (1 pt) & benzyl acetate (39 pts)	Alpha pinene, cineole, & benzyl acetate	Alpha pinene, beta pinene, cineole, & benzyl acetate
Euglossa asarophora	4		2
E. dodsoni	15	4	1
E. dressleri	2		
E. igniventris	1		
E. tridentata	1		
E. sp. (RD 355)	1		
E. sp. (RD 847)			1
Eulaema cingulata	7		
E. meriana	18	2	3
	49	6	7
Total bees—62			

[1] Tests were made at Cerro Jefe, Cerro Campana, and Santa Rita Ridge in central Panama in late January 1968.

Table 6. RESULTS OF FIELD TESTS OF VARIOUS PURE COMPOUNDS AND MIXTURES OF COMPOUNDS

	Cineole	Methyl salicylate	Benzyl acetate	Methyl benzoate	d-carvone	Linalool	2-phenylethanol	Alpha ionone	Beta ionone	Cineole (1 pt) & benzyl acetate (39 pts)	Alpha & beta pinene, cineole, & benzyl acetate	Alpha pinene, cineole, & benzyl acetate
Number of species of bees attracted	35	11	6	2	1	1	1	1	5	8	4	2

Total number of species of euglossine bees attracted 42
Total species of euglossine bees known from central Panama 57

Cineole proved to be a remarkable general attractant, bringing in 433 male euglossine bees (and no other kinds) representing 35 species. Methyl salicylate attracted 113 individuals of 11 species, while benzyl acetate attracted 36 individuals of 6 species. When 1 part cineole and 39 parts benzyl acetate were mixed (in approximately the proportions that they occur in the fragrance of *Stanhopea tricornis*), only 49 individuals of 8 species of bees were attracted. When alpha pinene was added in the proportion found in the same orchid, only six individuals of two species were attracted. One of those two species was *Eulaema meriana,* the known pollinator of *Stanhopea tricornis* in its natural habitat in Colombia and Ecuador. The other bees attracted were *Euglossa dodsoni,* a bee much too small to effectively pollinate *S. tricornis.*

FIGURE 7. Male *Eulaema meriana* visiting *Stanhopea tricornis.*

The bees behaved in quite the same manner when attracted to the liquid-soaked blotters as they do when attracted to flowers. They hovered in front testing the fragrance, landed and rubbed the blotters, launched into flight, and rubbed the tarsal pads on the tibeal scars. They also lost their wariness.

The attraction elicited by cineole, benzyl acetate, and methyl salicylate when tested in pure form indicates that these are general attractants. Cineole certainly was effective in attracting the majority of the species of euglossine bees in central Panama. Cineole occurs in approximately 60 percent of the euglossine-orchids analyzed. Benzyl acetate occurs in about 25 percent of the orchid species sampled and attracted fewer species of bees (about 10%). Methyl salicylate occurs in about 4 percent of the orchids sampled but attracted about 20 percent of the central Panama euglossine bees. Most of the species attracted to benzyl acetate or methyl salicylate were also attracted to cineole. The notable exceptions were *Eulaema bombiformis, Euplusia* sp. (RD 237), and *Eufriesia pulchra* which were attracted to methyl salicylate and not to cineole and *Eulaema cingulata* which was attracted to benzyl acetate but not to cineole or methyl salicylate. *Eulaema cingulata* is one of the most frequent euglossine-orchid pollinators and can be very selectively attracted to certain fragrances, for example, that of *Cycnoches ventricosum* (Dodson, 1965a). *Euglossa cyanura* was attracted only to 2-phenylethanol. Several species were attracted to beta ionone (see Table 4).

The mixtures of cineole and benzyl acetate with relative proportions similar to those found in orchid flowers (*Stanhopea tricornis* and others) did not attract the large numbers of species attracted by cineole or benzyl acetate alone (Figure 8). Clearly, the combination of the two attractants acts as a repellent for some species and modifies the attraction potential of the total fragrance. Alpha pinene does not act as an attractant when tested alone. When added to the mixture of cineole and benzyl acetate, a-pinene reduces the number of species attracted; therefore, a-pinene appears to act as a partial repellent or a modifier (Figure 8).

Most orchid species produce several components in addition to (or in place of) cineole, alpha and beta pinene, and benzyl acetate. We have not yet identified most of those compounds, but, if they follow the pattern established by the results cited above, some may prove to be general attractants while others may act as specific attractants as does 2-phenylethanol for *Euglossa cyanura.* Some probably modify the total attraction spectrum as does alpha pinene. The differential production of fragrance components by different species would thereby act to attract a very limited number of bee species even when numerous bee species occur in the habitat with the orchid.

Once the number of bee species attracted to an orchid species has been reduced, mechanical isolation can play a more effective role. Flowers may adapt to a small bee and thereby eliminate large bees as effective pollinators, or vice versa. Positioning of pollinaria can also

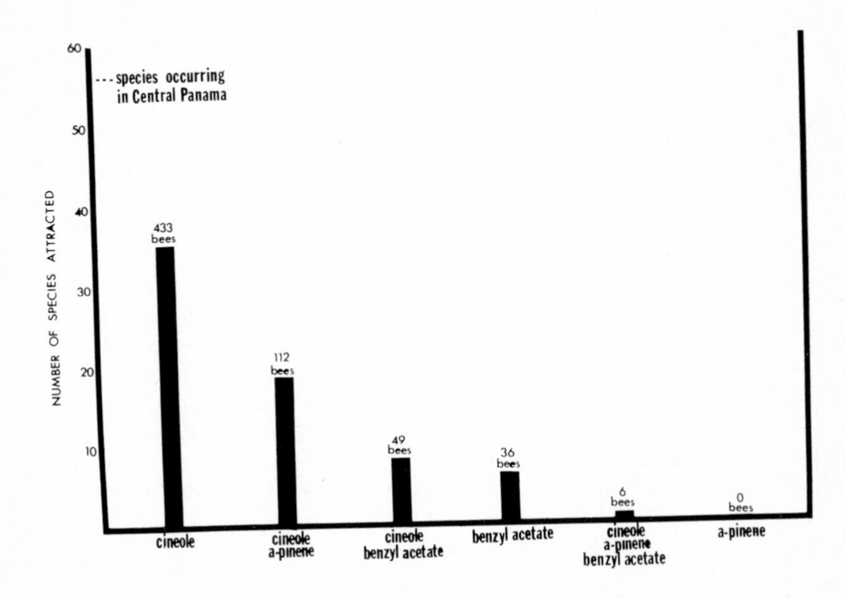

FIGURE 8. Graph showing the number of species of euglossine bees attracted to pure compounds and mixtures of compounds.

be effective at this stage, but this factor would generally lead to extensive modification of the floral apparatus with consequent recognition of the variants as distinct genera.

Most of the compounds identified at this point are relatively common natural plant products (Scora, 1967; Mooney and Emboden, 1968). Most are terpenes or aromatic hydrocarbons of low carbon number (8-10).

It has been pointed out that the euglossine-orchids produce attractants in the fragrance which attract specific kinds of bees. Recent work shows that the bees store the fragrance components in their hind tibeae (Adams, in press). The question immediately arises, Why should the bees do this? This question cannot be answered at this time. We plan to present carbon-14 labeled compounds to bees and attempt to trace the compound through the tibeae after transferal. It is possible that the compounds are metabolized by the bee or converted into other compounds, possibly pheromones for the attraction of the female. If this approach is feasible, we may be able to answer the question of why the bees collect the fragrance components.

The potential role of easily modified fragrances in orchid speciation

Speciation in euglossine-orchids can be easily facilitated by minor genetic changes affecting the production of fragrance components. The production of additional components or the failure to produce a component could change the attraction potential of the fragrance of a new genotype. If a different species of bee is attracted and the pollinator of the parental form is not attracted, a new reproductively isolated genotype could develop (Manning, 1957). In several instances we have found that different populations have slightly different odor spectra. For example, the members of the population of *Catasetum warscewiczii* at El Valle, Panama, have a slightly different series of components from the population at Cerro Campana. Both are probably pollinated by the same bee (*Eulaema bombiformis* pollinates this species in southern Costa Rica and in central Panama east of the canal). Further fragrance changes could attract different species of bees and isolate the two genotypes reproductively.

This system permits sympatric speciation; in fact it can make it quite simple. An example of this type may be occurring in the genus *Gongora* in Central America. Extensive populations of *Gongora "quinquenervis"* occur in Costa Rica and Panama. The plants are essentially the same in all external characters. The flowers tend to be variable in color within a given population, and in some cases morphological characters of the lip of the flower may vary. Early taxonomic treatments separated several forms on the basis of color. More recently, however, the tendency of taxonomists has been to reduce all gongoras which even resemble *G. quinquenervis* to one species. At Guapiles in Costa Rica, the large population of *Gongora* produces flowers which are structurally quite similar, but color of flowers, fragrance spectra, and pollinators diverge. One form (*G. unicolor*) produces flesh-pink flowers and a sweet fragrance (which we are unable to record on the chromatograph). It is recognizable only by the color of the flowers and the fragrance it produces, yet it is pollinated by a bee (*Euglossa purpurea*) which does not visit any of the other members of the group. The other members of the group produce flowers which are indistinguishable in form and vary widely in color from white flecked with red to completely wine-red. Different fragrance spectra are produced by certain members of this population. Two fragrance types have been identified. The different fragrance types, associated with pollinators, were designated "Guapiles No. 3 and No. 4"; "Guapiles No. 3" attracted *Euglossa gorgonensis,* while "Guapiles No. 4" attracted *Euglossa bursigera.* The significant point here is that only one of the three types of *Gongora* can be distinguished (and then only on the basis of color of flower) when pollinator data and fragrance spectra are ignored. The indication is that adaptive radiation to different pollinators is taking place. One form, *G. unicolor,* has reached a point of color differentiation. With time, the other two forms could develop differences so that they would be recognizable.

Other orchids

Considerable space has been devoted to a discussion of orchid speciation based on attraction of specific kinds of euglossine bees. These orchids constitute about 10 percent of the species in the family. What about the other 90 percent? How do they accomplish speciation on a comparative level? We know, for example, that numerous orchid species are adapted to pollination by moths. The syndrome of morphological adaptation to night-flying moths has been outlined by van der Pijl and Dodson (1966). This syndrome is characterized by the production of white or light green flowers, long nectar tubes, and strong fragrances at night. Because of the considerable difficulties in observing night pollination, little data have been accumulated concerning pollinator specificity in these orchids. Several genetically compatible species of a given genus are known to be sympatric in distribution. For example, the genera *Brassavola, Epidendrum, Habenaria,* and *Campylocentrum* in the Neotropics and the genera *Angraecum, Aerangis,* and *Platanthera* in the Old World tropics contain compatible sympatric species within each genus. Most of these genera consist of numerous species some of which are sympatric in each case, but few reports of natural hybrids exist. Preliminary studies with gas chromatography of members of the genus *Brassavola* indicate that quite distinct fragrance spectra exist (N. Williams, unpub.).

Similar situations appear to exist with such fly-pollinated genera as *Pleurothallis* in the Neotropics and *Bulbophyllum* in the Old World tropics. Both genera consist of more than a thousand species and have allied genera with numerous species. Scattered observations indicate that the species are isolated by pollinator specificity (Dodson, 1965; Ridley, 1890). Further study may demonstrate that the same kinds of phenomena are involved in the attraction of specific pollinators based on differential fragrance spectra.

Based on our pollination spectrum for the family (Table 1) about 33 percent of the members of the orchid family would be covered by the euglossine, moth, and fly-flower classes. Obviously, specific attraction by fragrance is not the only factor in speciation in the family; however, it is interesting to note that the majority of the large genera (those with over 1,000 species) are covered here. The one other large genus, *Dendrobium,* occurs through the island chains of tropical Asia where spatial isolation may have played a more important role.

Reciprocal Speciation of Orchid and Pollinator

This author has uncovered no evidence that the orchids have affected the evolution of their pollinators. On the contrary, the relationship between the orchid and its pollinator appears to be a one-way function. In most cases, the pollinator benefits in one manner or another from visiting orchid flowers. Most common is the presentation of food in the form of nectar to reward the visitor. As we have seen, a substantial number of orchid species produce fragrance components which are collected by male euglossine bees. It might be argued that the presence of the orchids as a source of these fragrance components would affect the evolution of the bees; however, these compounds are fairly common natural plant products, and the bees do visit and collect from other flowers which produce the compounds, such as anthuriums, gesneriads, and lecythids, as well as tree bark, roots, and wood (Figure 9). Even the classic example cited by Darwin, of *Angraecum sesquipedale,* with a ten-inch nectar spur, and the moth *Xanthopan morgani*

FIGURE 9. Male *Euplusia purpurata* scratching at wood.

var. *praedicta,* its assumed pollinator, is not a valid case of coevolution. The long-tongued moth has never been observed pollinating the orchid, but even if it does, the orchid is probably not the primary source of food for the insect. Other flowering plants with long nectar tubes exist in the environment and were probably on the scene long before the orchid. Consequently, it seems reasonable to assume that the moth adapted to other flowers with long nectar tubes (and vice versa) and that the orchid at a later time adapted to the long-tongued moth.

Literature Cited

Adams, R. M. 1966. Attraction of bees to orchids. Bull. Fairchild Trop. Gard. *21:* 6-13.

Allen, P. A. 1951. Pollination of *Coryanthes speciosa.* Bull. Am. Orchid Soc. *19:* 528-536.

Allen, P. A. 1952. The swan orchids, a revision of the genus *Cycnoches.* Orchid J. *1:* 226-229.

Allen, P. A. 1954. Pollination in *Gongora maculata.* Ceiba *4:* 121-124.

Cruger, H. 1865. A few notes on the fecundation of orchids and their morphology. J. Linn. Soc. London, Botan. *8:* 129-135.

Cruz Landim, C. da, A. Stort, M. da Costra Cruz, and E. Kitajima. 1965. Orgao tibial dos machos do Euglossini. Estudo ao microscopio optico e electronico. Rev. Brasil. *25:* 323-342.

Dodson, C. H. 1962. The importance of pollination in the evolution of the orchids of tropical America. Bull. Am. Orchid Soc. *31:* 525-534, 641-649, 731-735.

Dodson, C. H. 1965a. *Agentes de polinizacion y su influencia sobre la evolucion en la familia orquidacea.* Univ. Nac. Amazonia Peruana.

Dodson, C. H. 1965b. Studies in orchid pollination: The genus *Coryanthes.* Bull. Am. Orchid Soc. *34:* 680-687.

Dodson, C. H. 1966. Ethology of some euglossine bees. J. Kansas Entomol. Soc. *39:* 607-629.

Dodson, C. H., and G. P. Frymire. 1961. Natural pollination of orchids. Bull. Missouri Botan. Gard. *49:* 133-139.

Dodson, C. H., and H. G. Hills. 1966. Gas chromatography of orchid fragrances. Bull. Am. Orchid Soc. *35:* 720-725.

Dressler, R. L. 1968a. Pollination by euglossine bees. Evolution *22:* 202-210.

Dressler, R. L. 1968b. Observations on orchids and euglossine bees in Panama and Costa Rica. Rev. Biol. Trop. *15:* 143-183.

Dressler, R. L., and C. H. Dodson. 1960. Classification and phylogeny in the Orchidaceae. Ann. Missouri Botan. Gard. *47:* 25-68.

Ducke, A. 1902. As especies Paranenses do genero *Euglossa* Latr. Bol. Mus. Paranense. *3:* 1-19.

Manning, H. 1957. Some evolutionary aspects of the flower constancy of bees. Proc. Roy. Phys. Soc. *25:* 67-71.

Mooney, H. A., and W. A. Emboden. 1968. The relationship of terpene composition, morphology, and distribution of populations of *Bursera microphylla* (Burseraceae). Brittonia *20:* 44-51.

Moure, J. S. 1950. Contribuicao para o conhecimento do genero *Eulaema* Lepeletier (Hymenoptera-Apoidea). Dusenia *1:* 181-200.

Pijl, L. van der, and C. H. Dodson. 1966. *Orchid Flowers, Their Pollination and Evolution.* Univ. of Miami Press.

Porsch, O. 1955. Zür Biologie der *Catasetum*-Blute. Österr. Botan. Z. *102:* 117-157.

Roberts, R., and C. H. Dodson. 1967. Nesting biology of two communal bees, *Euglossa imperialis* and *Euglossa ignita* (Hymentoptera; Apidae), including description of larvae. Ann. Entomol. Soc. Am. *60:* 1007-1014.

Sakagami, S. F. 1965. Über den bau der männlichen Hinterschiene von *Eulaema nigrita* Lepeletier (Hymenoptera: Apidae). Sonderdruck aus Zool. Anz. *175:* 347-354.

Scora, R. W. 1967. Study of the essential leaf oils of the genus *Monarda* (Labiatae). Am. J. Botan. *54:* 446-452.

Vogel, S. 1963. Das sexuelle Anlochungsprinzip der Catasetinen- und Stanhopeen-Blute und die wahre Funktion ihres sogenannten Futtergewebes. Österr. Botan. Z. *110:* 308-337.

Vogel, S. 1966. Parfumsammelnde Bienen als Bestauber von Orchidaceen und *Gloxinia.* Österr. Botan. Z. *113:* 302-361.

Vogel, S. 1966. Scent organs of orchid flowers and their relationship to insect pollination. Proc. 5th World Orchid Conf., pp. 253-259.

Willis, J. C. 1966. *A Dictionary of the Flowering Plants and Ferns,* 7th ed., revised by H. K. Airy-Shaw. Cambridge Univ. Press.

Twenty-ninth Annual Biology Colloquium

Theme: Biochemical Coevolution

Dates: April 26-27, 1968

Place: Oregon State University, Corvallis, Oregon

Standing Committee for the Biology Colloquia: Paul O. Ritcher, *chairman,* Malcolm E. Corden, Ernest J. Dornfeld, Paul R. Elliker, Gwil O. Evans, Henry P. Hansen, Hugh F. Jeffrey, J. Kenneth Munford, John M. Ward

Special Committee for the 1968 Biology Colloquium: Kenton L. Chambers, *chairman,* William C. Denison, *co-chairman,* Derek J. Baisted, Robert Pacha, William P. Stephen, Henry Van Dyke

Colloquium Speakers, 1968:

Paul R. Ehrlich, professor of biology, Stanford University, *leader*

Cornelius H. Muller, professor of botany, University of California, Santa Barbara

Stephen J. Karakashian, associate professor, State University of New York, College at Old Westbury

Peter R. Atsatt, assistant professor, Department of Population and Environmental Biology, University of California, Irvine

Robert W. Hull, chairman, Department of Biological Sciences, Florida State University, Tallahassee

Lincoln Pierson Brower, professor of biology, Amherst College, Amherst, Massachusetts

Calaway H. Dodson, associate professor of biology, University of Miami, Coral Gables, Florida

Cooperating organizations:

National Science Foundation
School of Science, Oregon State University
School of Agriculture, Oregon State University
Honor Society of Phi Kappa Phi
Sigma Xi
Omicron Nu

Index